T0212579

SpringerBriefs in Agriculture

More information about this series at http://www.springer.com/series/10183

Muhammad Aslam · Muhammad Amir Maqbool
Rahime Cengiz

Drought Stress in Maize (*Zea mays* L.)

Effects, Resistance Mechanisms, Global
Achievements and Biological Strategies
for Improvement

 Springer

Muhammad Aslam
Plant Breeding and Genetics
University of Agriculture
Faisalabad
Pakistan

Muhammad Amir Maqbool
Plant Breeding and Genetics
University of Agriculture
Faisalabad
Pakistan

Rahime Cengiz
Maize Research Station
Sakarya
Turkey

ISSN 2211-808X   ISSN 2211-8098   (electronic)
SpringerBriefs in Agriculture
ISBN 978-3-319-25440-1   ISBN 978-3-319-25442-5   (eBook)
DOI 10.1007/978-3-319-25442-5

Library of Congress Control Number: 2015952763

Springer Cham Heidelberg New York Dordrecht London

Printed on acid-free paper

Springer International Publishing AG Switzerland is part of Springer Science+Business Media
(www.springer.com)

Contents

1 Introduction 1

2 Effects of Drought on Maize 5
  2.1 Effects on Crop Stand Establishment 6
  2.2 Effects on Growth and Development 8
  2.3 Effects on Reproductive Growth Stages 11
    2.3.1 Pollen Development 12
    2.3.2 Silk Development 13
    2.3.3 Pollination 14
    2.3.4 Embryo Development 14
    2.3.5 Endosperm Development 15
    2.3.6 Grain or Kernel Development 16

3 Mechanisms of Drought Resistance 19
  3.1 Drought Escape 21
  3.2 Drought Avoidance 22
  3.3 Drought Tolerance 23
    3.3.1 Osmotic Adjustment 24
    3.3.2 Antioxidative Defense Mechanism 26
    3.3.3 Plant Growth Regulators 28
    3.3.4 Molecular Mechanisms of Drought Tolerance 30

4 Global Achievements in Drought Tolerance of Maize 37
  4.1 Contribution of CIMMYT, IITA, and Other
     Collaborative Partners 38
  4.2 Contribution of Multinational Seed Companies 42

5   Biological Practices for Improvement of Maize Performance   45
      5.1   Screening for Drought-Tolerant Maize Germplasm.   45
      5.2   Conventional Breeding Strategies.   47
      5.3   Marker-Assisted and Genomic-Assisted Breeding   50
      5.4   Transgenic Maize Development.   54

6   Conclusions and Summary   57

References   59

Abstract

Drought is one of the most detrimental abiotic stresses across the world which is seriously hampering the productivity of agricultural crops. Maize is among the leading cereal crops in world, but it is sensitive to drought. Maize is affected by drought at different growth stages in different regions. Germination potential, seedling growth, seedling stand establishment, overall growth and development, pollen development, silk development, anthesis–silking interval, pollination, embryo development, endosperm development, and kernel development are the events in the life of maize crop which are seriously hampered by drought stress. Plants have developed numerous strategies which enabled them to cope with drought stress. Maize germplasms also have numerous features which enable some accessions to cope with drought stress in better ways. One of the adaptive strategies is the earliness which helps the plants to escape the drought stress. Some genotypes have the ability to avoid the drought stress either by reducing water losses or by increasing the water uptake. Drought tolerance is also an adaptive strategy which enables the crop plants to maintain the normal physiological processes and harbors higher economical yield under prevailing drought stress. Osmotic adjustment by accumulation of osmolytes, plant self-defense by accumulation of antioxidants, plant growth regulators, stress proteins, and water channel proteins, transcription factors and signal transduction pathways are involved in conferring the drought tolerance on maize. Maize genotypes have the differential capability to escape, avoid, or tolerate the drought stress. Great efforts were made by CIMMYT, IITA, and other multinational companies in the development of drought-tolerant maize open-pollinated varieties (OPVs) and hybrids. Drought-tolerant hybrids and OPVs of maize are being cultivated in numerous African countries. There is a need to further improve the level of adaptability against drought stress for combating the global issue of food security. Screening of available maize germplasms for drought tolerance, conventional breeding strategies, marker-assisted and genomic-assisted breeding, and development of transgenic maize are the prominent biological strategies for the improvement of encounterability against drought stress. Complete knowledge about the effects of drought stress, achievements, adaptive strategies, and possible breeding tools are the prerequisites for any breeding plan. Hence, these aspects were compiled in this brief.

Keywords Germination · Seedlings · Morphological traits · Physiological traits · Drought escape · Drought avoidance · Drought tolerance · Reproductive growth stage · Molecular factors

Chapter 1
Introduction

Biotic and abiotic factors of environment had modified the primal living organisms in such a way that modern plants were evolved. Any imbalance in these biotic and abiotic factors act as stress and impose serious hampering effects on growth and development of crop plants (Aslam et al. 2013a, b, c, 2014a, b; Naveed et al. 2014). Survival of the fittest enabled the superiorly evolving plants to prevail in the environment whereas, rest of the plants were eliminated during evolutionary events due to their fitness issues. Still there are lots of factors which continually reshaping the plant evolution; water availability is one of them. Water is among the basic needs of living organisms on this globe. Without any exaggeration, water availability modulated the water responsive signaling which acted as critical factor for shaping the flora on globe throughout the course of evolution. Different crops have different delta of water (*Box-1*) required for their usual growth and development. Maize required 368 kg, sorghum required 332 kg, barley required 434 kg and wheat required 514 kg of water for 1 kg dry matter accumulation (Rana and Rao 2000). Total 350–450 mm water is used by maize during its life cycle for completion of growth and development. Every millimeter of water is responsible for production of 10–16 kg grains and single maize plant consumes 250 litres of water at maturity (Du Plessis 2003).

Drought, salinity and low temperature stresses are known as main agricultural problems which are seriously inhibiting the plants to show off their genetic potential (Zhu 2002). Drought is sole factor which is affecting agricultural crops more than any other stress and it is becoming even more severe in different regions of world (Passioura 1996; Passioura 2007). Statistical data showed that globally area subjected to drought stress has doubled from 1970 to 2000 (Isendahl and Schmidt 2006).

Maize is major cereal crop of the world which is currently grown in large number of countries. It is a multidisciplinary crop and used in human food, animal feed, fodder and bioenergy production. Area, yield and production of maize across the world in comparison with rice and wheat have been shown in Figs. 1.1 and 1.2. Comparison showed that wheat is cultivated on more area relative to other cereals but yield and production of maize is more than wheat and rice (Figs. 1.1 and 1.2; FAOSTAT 2013).

© The Author(s) 2015
M. Aslam et al., *Drought Stress in Maize (Zea mays L.)*,
SpringerBriefs in Agriculture, DOI 10.1007/978-3-319-25442-5_1

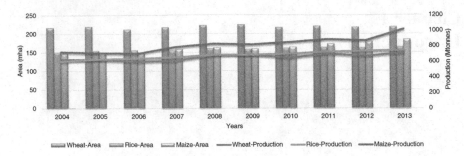

Fig. 1.1   Area (Mha) and production (MTonnes) of wheat, rice and maize across the world for last ten years (FAOSTAT 2013)

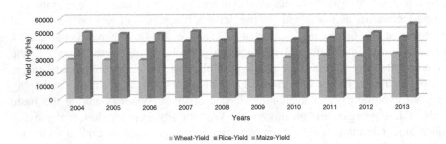

Fig. 1.2   Yield (Hg/Ha) of wheat, rice and maize across the world for last ten years (FAOSTAT 2013)

Drought stress is seriously affecting the maize crop resultantly hindering the productivity like other crops (Tai et al. 2011). Being drought sensitive crop, maize is affected at each and every stage of growth and development by lesser moisture availability. Prevalence of drought at seedling stage causes poor crop stand and under extreme conditions can result in complete failure of seedling establishment (Zeid and Nermin 2001). Shutting down of plant metabolism followed by plant death due to stomatal closure and inhibited gaseous exchange occurs in response to prolonged moderate drought stress (Jaleel et al. 2007). In case of maize reproductive growth stage is comparatively more sensitive to drought stress and under severe drought prevalence barren ear production might be the result (Yang et al. 2004). Global importance of maize and side effects of drought on maize triggered the breeders to develop drought tolerance maize germplasm. Drought responsive traits and adaptive mechanisms must be known for the development of drought tolerant maize stock. Genetic diversity assessed on the basis of adaptive mechanism like drought escape, drought avoidance and drought tolerance is present in maize genotypes. So, in this book we have compiled the information regarding effects of drought on maize plant from germination to harvest maturity. Different strategic options for the improvement of maize performance under drought stress are also included in this write up.

Box-1: Brief description of technical terms used in text

Delta of water is amount of water required for particular crop in complete growing season.

Germination index is defined as cumulative daily total of germination over specific days (Timson 1965). Speed of germination, totality and their interaction is also described as germination index (Brown and Mayer 1988).

Germination rate is defined as number of seeds of particular variety likely to germinate over a specific period of time.

Germination velocity index is defined as daily germination counting for estimation of seedling vigor (Throneberry and Smith 1955).

Seed vigor is cumulative properties of seed which determine the quick and uniform emergence potential of seed and followed by potential seedling development under diverse field conditions (AOSA 2002).

Seed priming is defined as maintenance of seed hydration level so, that necessary metabolic activities needed for initiation of germination can start but radicle emergence is avoided. Germination is improved by seed priming treatment. Seed treatment with water before sowing to improve the germination is called hydro-priming.

Osmopriming: seed treatment with osmotic solutions before sowing to improve the germination and stress tolerance.

Chapter 2
Effects of Drought on Maize

Water is vitally needed for every organism in specified amount and any deficiency in that particular amount imposes the stressful conditions. Water requirement is variable across the tissues and across the growth stages of same species of crop plant and maize crop has no exception so, far. Assessment of optimum plant water requirements is prerequisite to determine the water deficiency in plants. Water requirement of maize crop is low at early growth stages then reaches on peak at reproductive growth stages and during terminal growth stages requirement of water again lowers down. During reproductive growth stage, 8–9 mm water is needed per day to single plant. Four weeks are most crucial regarding water requirement which includes two weeks before and two weeks after pollination. Pollination is most critical growth stage for water requirement and all leaves are kept unfolded and grain yield is also decided at this stage. Grain filling and soft dough formation are most sensitive to water deficiency, whereas, pre-tasseling and physiological maturity are relatively insensitive to water deficiency. Drought stress during vegetative growth stages especially during V1–V5, reduces growth rate, prolong vegetative growth stage and conversely duration of reproductive growth stage is reduced (Pannar 2012). Each millimeter of water produces 15.00 kg of kernels and total 450–600 mm is needed across the whole season (Du Plessis 2003). Total 250 l water is consumed by maize plant till maturity (Du Plessis 2003). Relative water contents, stomatal resistance, water potential, leaf temperature and transpiration rate maintain the plant water relation and any imbalance in these or any one of these traits disturb the plant water relation (Anjum et al. 2011b). Relative water contents determine the status of metabolic activities of the cell or tissue. During early leaf development, relative water contents of the leaves were higher and tend to decline towards maturity. Strong correlation is reported between relative water contents, water uptake and transpiration rate. Under drought stress, relative water contents and water potential is reduced, resultantly, leaf temperature is increased due to reduced transpirational cooling (Siddique et al. 2001). It can be easily perceived that plant water status is dependent on stomatal activity (Anjum et al. 2011b).

Transpiration ratio is described as number of water molecules lost in order to fix one molecule of carbon. Soybean, wheat and maize have 704, 613 and 388 transpiration ratio respectively which shows that maize is relatively efficient water user crop (Jensen 1973). Despite of being efficient water user maize is badly affected by

M. Aslam et al., *Drought Stress in Maize (Zea mays L.)*,
SpringerBriefs in Agriculture, DOI 10.1007/978-3-319-25442-5_2

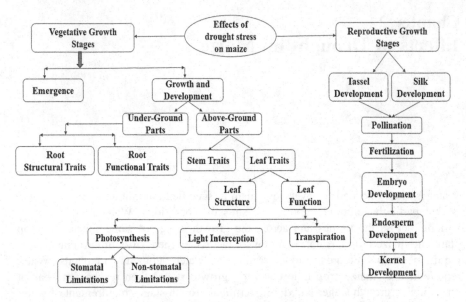

Fig. 2.1 Effects of drought stress on vegetative and reproductive growth stages of maize

drought stress due to hypersensitivity against water deficiency. In maize, developmental stages starting from germination to harvest maturity including seedling establishment, vegetative growth and development and reproductive growth stages are very much prone to drought stress. Effects of drought on maize at different growth stages and organizational levels have been presented in Fig. 2.1 and described in subsequent sessions.

2.1 Effects on Crop Stand Establishment

Crop stand establishment comprised of germination, emergence and seedling establishment. Concepts of germination and emergence prevailed under laboratory conditions and field conditions respectively. Crop establishment accomplished up to development of 7th or 8th leaf. These early growth stages are critical growth stages regarding drought stress. Always there are prominent differences among different levels of water treatments in maize regarding their effects at early growth stages (Fig. 2.2). Proper seed germination is dependent on availability of appropriate moisture contents for metabolic activation to breakdown the dormancy or to convert stored food into consumable form. Crop density or number of emerged seeds, mean time for emergence and synchronization of emergence are characteristic features which determined the efficacy of seedling establishment (Finch-Savage 1995). Crop survival, growth and development are determined by efficacy of seedling establishment (Hadas 2004). Drought stress reduces the germination potential of maize

Fig. 2.2 Maize seedlings subjected to different treatments of water deficiency. **a** 75 % of field capacity, **b** 50 % of field capacity, **c** 25 % of field capacity. Visual observations show that same set of maize genotypes is showing different pattern of growth and development due to differences in water availability. Leaf area, seedling height, stem girth and leaf rolling are clearly showing the significant differences among three different treatments. Leaf rolling seems highest in 25 % of field capacity whereas, leaf area, seedling height, stem girth are lower in this treatment relative to others. Under 75 % of field capacity, plant height, leaf area, stem girth seems to be greaterly higher than other two treatments

seeds by reducing their viability. Poor maize seed germination is directly associated with poor post germination performance (Radić et al. 2007). Severity of drought stress is directly linked with poor imbibition, germination and seedling establishment in maize (Achakzai 2009). Germination index (*Box-1*) is reduced by water deficiency (Almansouri et al. 2001). Germination potential, germination rate (*Box-1*) and seedling growth are the studied traits under drought stress because these traits are direct representative of crop establishment and are badly affected by drought stress (Delachiave and Pinho 2003). Germination velocity index (GVI) is corroborated with seed strength and always GVI (*Box-1*) was greater for maize hybrids than landraces due to hybrid vigor (Mabhaudhi 2009).

Maize grain size is greater than other cereals like wheat, rice and barley therefore, water requirement is greater for maintenance of osmotic potential and conversion of stored food into consumable form for proper germination (Gharoobi et al. 2012). Seed vigor (*Box-1*) is considered as important parameter in maize breeding which is badly reduced by drought stress (Khodarahmpour 2011). Water absorption, imbibition and metabolic enzymatic activation are hindered under limited water availability which reduces the maize grain germination. After germination, water deficiency significantly reduced the plumule and radicle growth which resulted in unusual seedling growth (Gharoobi et al. 2012). Hydropriming and osmopriming (*Box-1*) of maize seed result in improved seed germination by regulation of enzymatic activity to break the dormancy which clearly highlights the importance of water availability for exploitation of full germination potential (Janmohammadi et al. 2008). Root and shoot elongations are parameter of seedling growth and these are subjected to reduction by drought stress. At seedling stage in maize, reduction in shoot elongation is more than root elongation under drought stress (Khodarahmpour 2011). Seedling emergence rate of landraces is lower than hybrids whereas reduction in shoot elongation was less in landraces than hybrids under drought stress (Mabhaudhi 2009). Rate and degree of seedling establishment of maize are critical factors for determination of time of physiological maturity and grain yield (Rauf et al. 2007).

So, it is evident from above discussion that seed vigor, imbibition, germination potential, germination rate, plomule and radicle development and root and shoot growth of maize are adversely affected by drought at early growth stages.

2.2   Effects on Growth and Development

Proper growth and development of crop plants is important for establishment of normal plant structure that carry out all physiological and metabolic processes and give potential yield. Drought stress seriously hindered the growth and development of maize. Growth and development comprised of numerous component parameters which are estimated by different traits like, plant height, leaf area, structural and functional characters of root, plant biomass, plant fresh weight, plant dry weight and stem diameter. Plant height, stem diameter, plant biomass and leaf area are reduced under drought stress (Khan et al. 2001; Zhao et al. 2006).

Growth is described as increase in size of plant which is directly associated with increase in number of cells and cell size. Meristematic tissues are involved in active elongation of plant by active cell division. Cell division and cell size are reduced by reduction in water potential of cells which causes the reduction in plant growth (Nonami 1998).

Leaves in maize are ranged from 8 to 20 and these are present alternatively on nodes. Leaf is comprised of structural and functional components. Leaf growth consists of leaf size and number of leaves which are structural components. Photosynthesis, transpiration and light interception are the functional traits of leaf. Leaf size and number of leaves are reduced in maize by drought stress. Turgor pressure, light interception and flux assimilation are determinant of leaf elongation (Rucker et al. 1995). Wedge shaped motor cells are present on the upper leaf surface and these keep the leaves unfold whereas, under drought stress turgor of leaves is reduced and leaves are curled or folded (Du Plessis 2003). Leaf folding reduces the leaf area and resultantly light interception is reduced which decreases the photosynthetic activity. Leaf area and photosynthesis are directly proportional to each other (Stoskopf 1981). Cell division and cell elongation are reduced under drought stress which reduces the leaf area. Reduction in leaf area under drought stress conditions is taken as adaptive strategy by maize plants. Leaf area index is considered as an important parameter for maize breeding against drought stress (Hajibabaee et al. 2012). Plant water requirement is reduced by reducing the leaf area and probability of plant survival is increased under limited water availability (Belaygue et al. 1996) but chlorophyll contents, chloroplast contents and photo-synthetic activity are reduced which reduced the grain yield (Flagella et al. 2002; Goksoy et al. 2004).

Kinases protein family and cyclin-dependent kinases (CDKs) are involved in the active progression of cell cycle. CDK activity is reduced under water deficit conditions which increased the duration of cell division and decrease the number of cell divisions per unit time that ultimately reduces the growth of leaves and plant (Granier et al. 2000). Cell elongation is found to be reduced across all points on leaf. Common regulatory pathway is involved in cell division and cell elongation (Tardieu et al. 2000). Drought stress increases the leaf to stem ratio which is indication of high level of growth retardation in stems than leaves (Hajibabaee et al. 2012). Reduced water potential in roots interrupts the optimal water supply to the elongating cells and resultantly cell elongation is reduced. Water potential less than -10.0 Bars causes the reduction in leaf growth (Tanguilig et al. 1987).

Light interception is reduced after reduction of leaf area. Less interception of solar radiations causes the reduction in biomass production (Delfine et al. 2001). Besides light interception, stomatal activity is also responsible for lower biomass production (Delfine et al. 2001; Medrano et al. 2002). Rise in leaf temperature under drought stress, inhibits the enzymatic activity and reduces photosynthesis (Chaves et al. 2002). Photosynthetic machinery is inactivated by increase in leaf temperature above threshold temperature which is 30 °C (Crafts-Brandner and Salvucci 2002). Stomatal closure, reduced transpiration and its homeostatic effects are the cause of rise in leaf temperature under limited water availability (Jones 1992).

Photosynthetic activity in maize plant is reduced by stomatal and non-stomatal limiting factors. Reduced leaf turgor and root originated signals along with lower plant water status trigger the stomatal closure. Reduction of water potential in the roots transduces the signals for stomatal closure. CO_2 diffusion in the leaves is reduced by stomatal closure and supply of CO_2 to the RUBISCO is hampered (Flexas et al. 2007). Reduced CO_2 diffusion is considered as main reason for decline of photosynthesis. Abscisic acid (ABA) accumulation is increased in the leaves in response to drought induced signals which triggers the stomatal closure (Wilkinson and Davies 2010). Cellular environment becomes alkaline under drought stress. Rise in cellular pH increases ABA accumulation in the leaves (ABA trapping) which induced the stomatal closure (Jia and Davies 2007).

Stomatal closure has protective role in saving the water loss and increasing water use efficiency under mild drought stress but under severe drought stress stomatal closure becomes inevitable evil (Chaves et al. 2009). Stomatal conductance and transpiration rate modulate the CO_2 diffusion in leaves which are directly linked with stomatal opening. CO_2 fixation rate, intercellular CO_2 concentration and net photosynthetic rate are the parameters used for assessment of stomatal conductance and photosynthetic activity under drought stress (Sage and Zhu 2011). Passive and active stomata closures occur under normal conditions and stress prevalence respectively (Fig. 2.3). Different genes are regulated to maintain the production and consumption equilibrium by alteration of redox state in leaves under drought stress. Reactive oxygen species (ROS), electron acceptors and electron carriers have potential role in regulation of stomatal conductance (Chaves et al. 2009).

Leaf structural characters and biochemical parameters are components of non-stomatal inhibition of photosynthesis. According to Von Caemmerer (2000)

Fig. 2.3 Passive and active stomatal closure. Passive stomatal closure occurs under normal conditions and active stomatal closure occurs under drought stress (Arve et al. 2011). © 2011 Arve LE, Torre S, Olsen JE, Tanino KK. Originally published in [short citation] under CC BY-NC-SA 3.0 license. Available from http://dx.doi.org/10.5772/24661

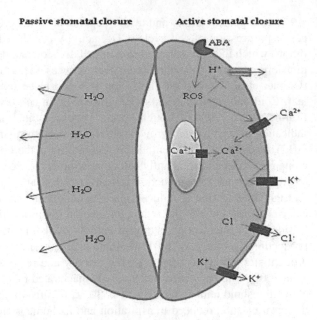

and Ghannoum (2009) carboxylation is changed by RUBISCO (Ribulose 1,5-bisphosphate carboxylase/oxygenase), PEPC (phosphoenolpyruvate carboxylase) and regeneration of PEP (phosphoenolpyruvate). Activity of the enzymes involved in the photosynthesis are reduced in case of non-stomatal inhibition of photosynthesis. Chlorophyll contents are reduced either by activation of cellular protein degradation or by limited nitrate synthesis (Becker and Fock 1986; Ghannoum 2009).

Maize is C4 plant and it is reported in C4 plants that intercellular spaces and chloroplast positions are misplaced by drought stress resultantly CO_2 diffusion and light penetration are disturbed followed by decreased photosynthetic activity (Flexas et al. 2004). Photorespiration and Mahler's reaction act as alternative electron sinks under drought stress (Ghannoum 2009). Mahler's reaction is involved in generation of reactive oxygen species and develops oxidative stress under drought stress. Oxygen molecule is converted into superoxide as a result of direct reduction reaction in Photosystem-I (Haupt-Herting and Fock 2002). Photosynthetic metabolism is reduced by reduction reaction of carbon substrate. Carboxylation activity of RUBISCO, regeneration of RuBP and ATP are reduced by inhibited CO_2 concentration in the leaves under drought stress (Tezara et al. 1999). CO_2 diffusion through mesophyll is reduced due to change in carbon metabolism and leaf photochemistry under drought stress. Leaf biochemistry, membrane permeability (aquaporin activity), leaf shrinkage, alterations in intercellular spaces, intercellular structure, internal diffusion and internal conductance are altered under drought stress which results in reduction of CO_2 diffusion through mesophyll (Lawlor and Cornic 2002; Chaves et al. 2009).

Roots have the critical importance for plant because these are the primary detectors or sensors of drought stress. Root length, root volume, root density and number of roots are the characteristic structural traits which are disturbed under drought stress and resultantly whole arial plant parts are disturbed. Spatial water uptake and temporal water uptake are functional traits of roots. Root system of maize comprised of axillary and lateral roots. Axillary roots are further comprised of primary, seminal, nodal or crown roots (Cahn et al. 1989). Primary and seminal roots are collectively known as embryonic roots. Seminal roots are permanent and have functional role in growth and development of plant (Navara et al. 1994). Roots of maize plant becomes elongated under mild drought stress to explore the more soil foils for more water uptake whereas, under severe drought stress root length is reduced. Root density, volume and number of roots are reduced under mild and severe drought stress (Nejad et al. 2010).

Requirement of photosynthates and energy is reduced in leaves due to reduced leaf area by leaf rolling or curling under mild drought stress. Photosynthetic assimilates from leaves are directed toward roots for their elongation to increase the water uptake (Taiz and Zeiger 2006). Roots act as primary sensor of water deficiency in soil and transduce signals to the aerial parts to modulate the growth and development. Signal from roots to the aerial parts are transduced through chemical and hydraulic vectors (Davies et al. 1994). Decreased water and nutrient uptake increase the pH of xylem (reduction of negative or positive ions) which transduces ABA-mediated signals to the leaves for preventing water loss by stomatal closure (Bahrun et al. 2002). Reduction in root growth under drought stress is also associated with reduced cell division and cell elongation. Microtubules are critical for cell division and cell elongation because these microtubules are involved in cellular morphogenesis, embryo development, organogenesis, stomatal conductance and organ twisting (Steinborn et al. 2002; Whittington et al. 2001; Marcus et al. 2001; Thitamadee et al. 2002). Reduced root turgor under dehydrated conditions, increases ABA accumulation and plasmolysis. Plasmolysis seriously damages the microtubule skeleton and cellular geometry (Pollock and Pickett-Heaps 2005). Disrupted microtubules in roots induce the ABA accumulation by increasing ABA biosynthesis. Interactions between microtubules, cell wall, plasma membrane and ABA biosynthesis are reported under osmotic stress (Lu et al. 2007).

2.3 Effects on Reproductive Growth Stages

Drought has adverse effects on maize life cycle; particularly reproductive growth phase is most susceptible to drought stress. Translocation of photosynthetic assimilates to the reproductive parts rather than roots for their extensive elongation is most probable reason for more susceptibility of maize plant during reproductive growth stage under drought stress (Setter et al. 2001; Taiz and Zeiger 2006). Sequential effects of drought stress on reproductive growth stages of maize are described in Fig. 2.1. Pollen and silk development, pollination, embryo development, endosperm development and kernel development are the different component phases of reproductive growth stage which are severely threatened by drought stress.

2.3.1 Pollen Development

Pollens are produced in the tassel which is present on top of the plant. Almost 25 million pollens are produced by single tassel under normal conditions. Modern hybrids produce 2–5 million pollens and landraces produce about 14–50 million pollens on an average (Burris 2001). Breeding efforts are always focused to reduce the tassel size and make tassel smart enough to ensure maximum photosynthetic reserves supply to female inflorescence rather than male inflorescence development because there is no problem of pollen availability in case of maize. Breeding efforts had suppressed the male over-dominance which had reduced the pollen production capability (Duvick and Cassman 1999). Maize pollens are produced in huge bulk and crop yield is not affected even due to 40 % reduction in pollen production (Du Plessis 2003). Timing for pollen shedding is effected negligibly by drought stress so, pollen shedding occurs mostly at normal time even under drought stress but in severe drought cases pollen shedding is adversely affected. Anthesis-silking interval is increased by decrease in silk growth and development rate. Pollen shedding depends on type of variety and environmental conditions; pollen shedding may continue from 2 to 14 days under normal conditions. About 1 h after sunrise is the time for initiation of pollen shedding which remain continue for 4–5 h and maximum pollen shedding occur during 5–8th day of pollen shedding (Burris 2001).

Pollens are affected by drought stress in different ways. Pollen mortality occurred due to dehydration as moisture of pollen is lost due to drying conditions (Aylor 2004). Settling speed, pollen viability, specific gravity, pollen shape and dispersal are seriously affected in dehydrated pollens (Aylor 2002). Increased ABA accumulation and reduced invertase activity are the main reasons for pollen sterility under drought stress (Saini and Westgate 2000). Conversion of sucrose to hexoses is impaired by reduced invertase activity (Sheoran and Saini 1996). Pollens of maize were studied under drought and high temperature stresses which showed that pollen weight, pollen viability, pollen size, pollen tube length and pollen moisture contents were affected by these stresses (Fig. 2.4).

Maize pollens are of large size as compared to other angiosperms and have relatively higher moisture contents. Pollen viability is reduced greatly if pollen moisture contents are reduced below 0.4 g per gram of pollens (Buitink et al. 1996). Pollens absorb moisture from hydrated silk to initiate proper germination so, pollen germination is reduced in case of dehydrated silk under drought stress (Heslop-Harrison 1979). Starch and certain osmolytes are present in the pollens which protect them from losing viability. Drought stress reduced the accumulation of starch in pollens during pollen development which rendered them nonfunctional (Schoper et al. 1987). Upregulation of galactinol and vacuole invertase genes in

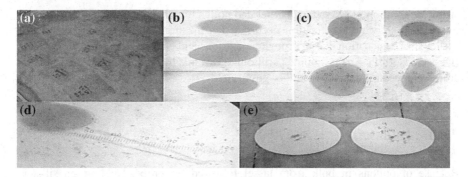

Fig. 2.4 Evaluation of different aspects of maize pollens under drought stress; **a** pollens collected from field and stored in zipper bags, **b** viable pollen grains; **c** measurement of pollen grain size; **d** pollen tube length; **e** pollen moisture contents measurement. These measurements showed that pollen viability, pollen size, pollen tube length and pollen moisture contents were reduced under drought stress

pollen under drought stress showed that these protect the pollens through osmo-protection and prevent the loss of viability (Taji et al. 2002). Gene expression is changed in such a way that cell wall structure and synthesis is impaired which results in loss of pollen viability under drought stress (Zhuang et al. 2007). Severe drought stress at tasseling stage reduce the yield by affecting the number of kernels per row, number of kernel rows, harvest index, number of kernels per cob and grain yield per plant (Anjum et al. 2011b). Increase in ABA accumulation up to 0.5 μM favor the pollen germination and pollen tube elongation but further increase in ABA contents significantly reduces the pollen germination and pollen tube elongation (Zhang et al. 2006).

2.3.2 Silk Development

Silk is female floral part of maize plant and should be receptive for proper polli-nation and fertilization. Silks remain receptive for 21 days but receptivity tends to reduce 10 days after silking (Du Plessis 2003). Silk elongation starts from the butt of the ear and terminal portion of cob elongated at the end. Large ear size delayed the silk appearance. However, it is reported that silking is delayed by 6–9 days by prevalence of drought stress (Dass et al. 2001). Tassel emerges 2–4 days earlier than silk emergence under normal conditions and this pattern is called protandry. Delay in appearance of silk under drought stress conditions is responsible for increased anthesis-silking interval (ASI) which is very critical index for efficient completion of reproductive growth stage. Lower the value of ASI higher will be the productivity and vice the versa. After silking, silk continue to elongate until it is pollinated and lengthwise it may reach up to 15 cm (Bassetti and Westgat 1993). After fertilization, elongation of silks stops and desiccation starts. Under drought stress, desiccation of silks starts earlier and pollen tube becomes unable to reach the ovary resultantly no fertilization occurred. Fertilization failure occurs because of earlier silks desiccation due to drought conditions and ear bareness becomes the fate (Dass et al. 2001). So, assimilate partitioning towards the silk and hydration of silks are of prime importance for higher grain yield.

2.3.3 Pollination

Release of pollens in bulk from tassel followed by proper landing on silks is necessary for successful pollination process. Losses due to pollination failure can never be recovered even after rehydration and yield losses may reach up to 100 % (Nielsen 2002). Pollen grain productivity reduces from 3 to 8 % on daily basis under drought stress (Rhoads and Bennett 1990). Pollen shedding is accelerated and silking is delayed by drought prevalence for four consecutive days and this increases the anthesis-silking interval followed by 40–50 % yield losses (Nielsen 2005a, b). Development of silk and ear is dependent on sufficient sugar supply which results in potential seed setting (Zinselmeier et al. 1999). Invertase (carbohydrate transporter) activity is reduced under drought stress which reduces the carbohydrate supply to the developing reproductive plant parts. Glucose contents in the pedicle of ovary are reduced due to IVR2 (soluble invertase) reduction during pre and post pollination under drought stress (Qin et al. 2004). Starch contents of the floral parts are reduced under drought stress due to impaired activity of the enzymes involved in starch metabolism (Zinselmeier et al. 2002). Pollination process is disturbed in following ways by drought stress; (a) silk becomes dried under dehydrated conditions and no more supportive for pollen tube development (Nielsen 2002), (b) pollen shedding occurs before silking which causes increase in anthesis silking interval (Nielsen 2002), (c) silk elongation rate is reduced (Lauer 2012), (d) silk becomes non-receptive for pollen grains under dehydrated conditions along with low humidity (Nielsen 2005a, b). So, the pollination process is badly affected by drought stress in maize causing low productivity at the end.

2.3.4 Embryo Development

Embryonic development is very susceptible to drought stress. During early embryonic development, embryo abortion occurs due to drought or heat stress (Setter et al. 2011). Drought stress prior to fertilization can cause embryo abortion (Andersen et al. 2002). Grain yield in maize is mainly dependent on the tolerance of female reproductive part. Reactive oxygen species are accumulated in the ovary as a result of drought stress and embryo is aborted in oxidative environment (Kakumanu et al. 2012). Embryo sac development is impaired due to imposition of drought stress during megaspore mother cell formation and resultantly 80–90 % yield losses are reported (Moss and Downe 1971). Insufficient provision of photosynthetic assimilates and sugar substrates to the developing embryo cause their abortion (Feng et al. 2011). Soluble invertases (Ivr2) and cell wall associated invertases are responsible for the provision of hexose to the developing embryos. These invertases are suppressed under drought stress causing check to supply of sugars and assimilate to embryo resulting embryo abortion (Andersen et al. 2002; Feng et al. 2011). Sucrose (substrate for invertase) to hexose ratio is very important for normal embryo development which is impaired during drought stress. Cell wall associated invertases and sugars are involved in signaling pathways and theses signaling pathways are affected by disturbance in expression of invertases and sugars (Kakumanu et al. 2012). Exogenous application of nutrients at reproductive stages rescue the 80 % embryos which proves that assimilate translocation is major reason for embryo abortion relative to lower water potential which causes comparatively less damage (Boyle et al. 1991). Leaves upload sucrose in phloem then it reach to pedicle where invertases hydrolyse sucrose into glucose and sucrose. These hexoses are used for kernel development (Cheng et al. 1996) and starch biosynthesis which participate in ovary development. ABA accumulation triggers the embryo abortion under drought stress (Setter et al. 2001). So, embryo development is very susceptible reproductive growth stage to drought stress which is affected by different ways.

2.3.5   Endosperm Development

Endosperm is storage house of food for embryo in the seed and like other reproductive stages; endosperm development is seriously affected by drought stress. Storage capacity of the endosperm is determined by cell division during early developmental stages of endosperm whereas; final volume of endosperm is determined by cellular elongation and multiplication of cellular organelles (Olsen et al. 1999). Cell division is reduced by imposition of drought stress during endosperm development and resultantly storage capacity is reduced (Ober et al. 1991). Prevalence of drought stress after fertilization, suppresses the cell elongation and multiplication of organelles causing reduction in final endosperm volume.

Process of endoreduplication occurs in the endosperm after mitotic cell division. Endoreduplication is repetition of S phase (synthesis phase) with mitotic cell division. There is no cytokinesis but DNA ploidy becomes double after every repetition of endoreduplication. Cell enlargement, cell differentiation, survival and metabolic activities are the key functions of endoreduplication (Barow and Meister 2003). Comparative evaluation showed that endoreduplication is less affected by drought relative to mitotic cell division (Artlip et al. 1995). Transition from mitotic cell division to endoreduplication is also affected by drought stress (Mambelli and Setter 1998). Cell division is reduced during early stages (1–10 days after pollination) of endosperm development in the apical kernels whereas; endoreduplication is reduced during terminal stages (9–15 days after pollination) of endosperm development (Setter and Flannigan 2001).

2.3.6   Grain or Kernel Development

Kernel development is very important phase as for as productivity is concerned and comprised of following component stages; blister stage, soft dough stage, milking stage, hard dough stage and dent stage. High moisture contents are needed during blister stage for grain filling and drought stress at this stage results in poor quality kernels. Moisture requirement during soft dough, milking and hard dough stages is higher enough that drought stress at these stages can reduce the kernel quality and yield. Drought stress during hard dough stage causes the premature hanging of the cobs. Water requirement of dent stage is lower relative to pre-dent stages of kernel development but drought stress at this stage still can cause potential loss in yield and quality (Pannar 2012).

Kernel development in maize is comprised of three major stages; (a) lag phase; sink capacity is developed, water contents increase and biomass accumulation reduces (Saini and Westgate 2000), (b) effective grain filling stage or linear phase; maximum biomass accumulation occurs in this stage and kernel size is determined (Westgate et al. 2004), (c) physiological maturity; maximum dry weight is gained and later on grain enters in quiescent phase (Saini and Westgate 2000).

Sink capacity and source strength interact with each other for grain filling. Differences in grain weight are due to difference in source sink ratio. Source strength is determined by photosynthesis and carbohydrate assimilation whereas, sink capacity is determined by sink's activity (Westgate et al. 2004; Yang et al. 2004). Drought stress reduces the photosynthesis and translocation of photosynthetic assimilates followed by reduced grain filling. Source strength and sink capacity are reduced by drought stress in maize. Grain size reduction is caused by reduced remobilization of photosynthetic assimilates (Yadav et al. 2004). Grain filling is also reduced due to decreased activity of sucrose and starch synthesizing enzymes under drought stress (Anjum et al. 2011b). Numbers of kernels are determined during pre-anthesis stages whereas; kernel weight is determined at post-anthesis stages. Drought stress during post-anthesis stages is responsible for kernel weight reduction (Oveysi et al. 2010). Interaction of water and biomass during kernel development are the determinants of final kernel volume. Water contents of the kernel are increased during early developmental stages of kernel and later on water contents decrease followed by increase in biomass accumulation. Biomass accumulation is dependent on source strength and sink's capacity which are seriously reduced by drought stress so final kernel volume is reduced by drought stress (Gambín et al. 2006). Reduced water potential and kernel water uptake squeeze the duration of kernel filling resultantly kernel size is reduced (Brenda et al. 2007). It is reported that drought stress during, kernel development is responsible for 20–30 % yield losses which are mainly due to under sized kernels (Heinigre 2000). Another report mentioned that drought prevalence during kernel development can cause 2.5–5.8 % yield losses on daily basis (Lauer 2003).

Chapter 3
Mechanisms of Drought Resistance

Effects of drought stress are very uncertain and unpredictable because they impair the yield, yield potential and across the years performance. However, selection of genotypes with better yield under drought prevailing conditions is effective tool for combating against drought stress. Heterogeneity in nature of drought stress, variable effects in space and time, degree and severity of stress are further increasing the erratic and unpredictable behavior of stress. Nature has bestowed the plants to adapt for survival and productivity under stressful conditions (Gill et al. 2003). Plants harbor different morphological, physiological and biochemical traits which enable them to adapt or resist under stressful conditions. Resistance can be described as least reduction in yield under drought stress conditions relative to normal water availability. Resistance can be in the form of escape, avoidance and tolerance (Bohnert et al. 1995). In evolutionary reference, drought resistance is described as ability of the varieties or species to survive and reproduce under limited water availability. In agricultural context, drought resistance is described as ability of the plants to produce economical yield under limited water availability (Qualset 1979).

Plant mechanisms which contribute to bring least losses in yield under drought prevailing conditions compared to higher yield under normal water availability are also described as drought resistance. Existence of genetic variability among different crop plants and varieties of same species for drought resistance was reported. Yield stability in wheat, maize, rice, barley and sorghum was used to determine the drought resistance (Singh 2010). Differences in yield, yield stability and level of drought resistance showed that improvement can be made by proper exploitation of this genetic variability. Different sources can be used for improvement of drought resistance in crop plants and some of them are as following; cultivated varieties, land races, wild relatives and development of transgenes. Landraces and wild relative are possible option to get the genes for drought resistance and to incorporate them in modern cultivars to improve their status of drought resistance. *Zea maxicana* or *tripsacum floridanum*, wild relative of maize, are tremendous source of novel genes for improvement of tolerance against drought and other stresses (Singh 2010). Transgenes can be developed if genes for drought resistance are available in non-crossable parents.

© The Author(s) 2015
M. Aslam et al., *Drought Stress in Maize (Zea mays L.)*,
SpringerBriefs in Agriculture, DOI 10.1007/978-3-319-25442-5_3

Existence of significant genetic variability among or between genetic populations is prerequisite for genetic improvement. Stress breeders suggested that elite
breeding populations have very low frequencies of stress resistant alleles so, first of
all these populations must be evaluated (Blum 1988). Significant genotypes × environment interaction (GEI) in response of drought stress showed that substantial
genetic variation exists in breeding population. Breeders recommend that selection
of breeding population for improvement is perquisite (Hallauer and Miranda 1988).
Most importantly genetic variation for kernel yield under stress and stress free
conditions is conducive. Elite breeding genotypes or cultivars have significant
variability for drought resistance related traits which should be used for improvement of drought resistance on priority basis. So yield is primary trait associated with
drought resistance whereas, appropriate secondary traits which confer drought
resistance are selected as selection criteria if following assumptions are full filled;
trait should be genetically correlated with yield, should have higher heritability,
should be of stable nature, easy to measure and traits should be correlated with yield
losses under normal prevailing conditions (Edmeades 2008; Barker et al. 2005).
Most of secondary traits do not full fill all of these prerequisite criteria however,
effective secondary maize traits associated with drought resistance are; leaf rolling,
stay green, shorter anthesis silking interval, cob barrenness (number of kernels per
ear), root system, increased leaf erectness, kernel weight and low canopy temperature (Bolaños and Edmeades 1996; Edmeades et al. 2000). Genetic variability in
secondary traits of maize i.e. yield components and physiological traits can also be
exploited to accelerate the improvement in yield under drought stress.

Numerous factors contributed to the increase in resistance of maize germplasm
against drought stress i.e. high plant density in field during development of inbred
lines, prevalence of drought and heat stress in nurseries with insufficient water
availability, use of high yielding and stable progenitors for breeding program and
multi-location testing of material (Duvick et al. 2004; Tollenaar and Lee 2011).
Major breeding objectives are to cultivate maize hybrids with greater yield
potential, stable yield and improved grain traits for user whereas an additional on
demand objective is to produce the hybrids with enhanced resistance against
adversaries. New genotypes of maize developed through keeping in view the above
mentioned objectives, would overcome the water deficiency by lowering the yield
penalty. This implies that all maize hybrids should have significant level of drought
resistance (Kitchen et al. 1999). Heterosis acts as important mechanism for stress
tolerance, as maize hybrids give higher yield even under drought stress relative to
maize varieties (Blum 1988). In maize, kernel yield is critically determined during
flowering and early grain development (Claassen and Shaw 1970; Shaw 1976).
Drought resistance is not heritable plant trait but numerous mechanisms are involved in conferring the resistance in different ways. These mechanisms are classified into three different types: drought escape, drought avoidance and drought
tolerance (Fig. 3.1).

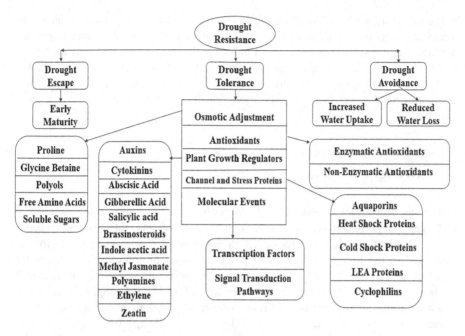

Fig. 3.1 Mechanism of drought resistance in maize with background traits like, morphological, physiological, biochemical and molecular traits

3.1 Drought Escape

In low land tropics, 400–500 mm is lower limit of rainfall for optimum maize cultivation, in mid tropics 350–450 mm and in highlands 300–400 mm. Water use efficiency (WUE) is lower in low lands so more rainfall in needed for proper crop growth than highlands. Synchronization of phenology with water availability is selection objective in breeding for earliness (Edmeades 2013). Shortening of growing season or life cycle of crop before the prevalence of drought stress is described as drought escape. Drought escape is important against terminal drought stress so, reproductive growth stage is to be involved for escaping drought (Araus et al. 2002). Days to sowing, days to flowering and days to maturity are genetically heritable traits and selection can re-modulated the phenology with water availability. Interaction between genotype and environment (G × E) is determinant of crop duration and induce the plants to complete the life cycle before the onset of drought. Successful synchronization of phenological development with availability of soil moisture and predominance of terminal drought stress are the factors which necessitate the adoption of drought escape (Araus et al. 2002). Development of short duration and early maturing cultivars is helpful for escaping the crop from terminal drought stress (Kumar and Abbo 2003). Yield is directly linked with duration of the crop and any reduction in crop duration ultimately reduces the yield (Turner et al. 2001). Crop growing season can be partitioned in two parts; sowing to flowering and flowering to physiological maturity. Early, intermediate and late maturing cultivars can be developed by making selection as these characters are heritable. These stages are heritable in nature and selection can be made easily for earliness as earliness is favorable for drought escape (Bänziger et al. 2000). Earliness is characteristic feature of drought escape whereas, early maturing genotypes have lower evapo-transpiration, lower leaf area index and lower yield potential.

3.2 Drought Avoidance

Numerous physiological and metabolic processes are not exposed to water stress and keep on performing normal functions in case of drought avoidance (Blum 1988). Drought avoidance is measured as estimate of tissue water status that is expressed in the form of turgor water potential under drought stress. Plant water status is maintained by either reducing transpiration rate or by increasing water uptake. Leaf rolling, leaf firing, canopy temperature, stomata closing, leaf attributes, and root traits are important selection criteria for drought avoidance in maize. Stomata are involved in transpiration and gaseous exchange (photosynthesis and respiration). Plant maintain water status by closing their stomata and avoiding the water losses (Turner et al. 2001; Kavar et al. 2007). How stomatal closure prevents the water losses under drought stress has been discussed in Sect. 2.2. However, stomata closure confers yield penalty due to critical effects on photosynthesis and respiration. Leaf rolling and leaf firing are important traits used for assessment of drought avoidance in maize. Insufficient leaf transpiration cause the dryness of leaves which results in leaf firing. There was report of negative correlation between leaf senescence and yield in maize. Canopy temperature is also negatively correlated with yield in maize under drought stress (Singh 2010; Edmeades 2013).

Maintenance of water uptake can be accomplished with the help of extensive root system (Turner et al. 2001; Kavar et al. 2007). Structural/architectural or phenotypic and functional or hydraulic root traits are very important for effective drought avoidance in maize. Length of apical zone, length of basal zone, length of branch inter-space, maximal number of branches, initial root elongation, insertion angle, root radius, standard deviation of the root tip heading and tropism type (geotropism for primary roots and exotropism for secondary roots) are used as architectural or phenotypic traits in maize for estimation of water uptake pattern with the help of mathematical models (Doussan et al. 1998; Leitner et al. 2014). Hydraulic root traits e.g. axial conductance between crowns of maize nodal roots, non-homogeneous soil water potential, transpiration during day time, water potentials in the root and hydraulic conductance are examined for modeling of water absorption and finding responses against water stress (Doussan et al. 1998; Leitner et al. 2014). These structural and functional models showed that higher water availability in root axis was responsible for higher transpiration. It is also reported that root hydraulic properties were responsible for determination of temporal dynamics of water deficit in root zone. Water saving and water spending behavior of roots is attributed to lower equivalent root conductance and higher root conductance respectively. Water uptake and root conductance is dependent on hydraulic properties of roots, age of roots and order arrangements of roots. It is found that radial conductivity of lateral roots is more influential than root structural traits on total transpiration (Leitner et al. 2014). So, this structural and functional root modeling in maize clarified that both architectural and hydraulic properties of roots are important for avoiding drought stress.

Thick and deep root system is conducive for extraction of more water from soil. Root characters like, root length, root density and root biomass are the main determinants of drought avoidance (Turner et al. 2001; Kavar et al. 2007). Thickness of individual root, maximum root depth (Ekanayake et al. 1985), root length over weight ratio (Aina and Fapohunda 1986), weight of seminal roots (Tuberosa et al. 2002a) and root length density (Aina and Fapohunda 1986) are the traits of roots which ensure the excessive water uptake under prevailing drought stress.

Pubescence or hairiness is very important character of xeromorphic plants. Reduction of leaf temperature and transpiration are the characteristic benefits of leaf hairs (Sandquist and Ehleringer 2003). Bicellular microhairs, prickle hairs and macrohairs are the three hair types reported in maize leaves. Macrohairs are prominently visible and act as morphological marker for identification of adult leaf (Moose et al. 2004). Glaucousness or waxiness of leaves is involved in maintaining the water potential of leaves (Ludlow and Muchow 1990). It is reported that temperature of glaucous leaves remains lower than non-glaucous leaves. It is claimed that 0.5 °C reduction in leaf temperature for six hours on daily basis can increase the grain filling time for three days (Richards et al. 1986). Two pathways are involved in the biosynthesis of wax in maize. One pathway biosynthesize the wax till 5th or 6th juvenile leaf and wax of this pathway consists of mainly alcohols and aldehydes. Second pathway biosynthesize the wax throughout the life and consists of mainly esters (Bianchi et al. 1985). About twenty GLOSSY genes (genes for cuticle wax production) are present in maize which determines the quantitative and qualitative features of wax (Neuffer et al. 1997). There is significant genetic variability observed in maize for glaucous versus non-glaucous and hairy versus non-hairy. So, there is strong consensus that stomatal conductance, hairiness and glaucousness/waxiness can reduce the water losses whereas; structural and functional characteristics of roots can improve the water uptake to enable maize plants to avoid drought stress.

3.3   Drought Tolerance

Potential of crop plants to maintain their growth and development under drought stress is termed as drought tolerance. Yield stability is also associated with drought tolerance under prevailing drought conditions. Tolerance is very complex mechanism and plants have evolved numerous adaptations at physiological and molecular levels to confer drought tolerance. Higher economic yield under drought stress is the characteristic feature of drought tolerant accessions. Survival is important at seedling stages whereas, later on just survival without economic yield have no importance for breeders and farmers (Bänziger et al. 2000). Plant growth and development, plant phenology, grain filling and translocation of photoassimilate reserves are important traits to be targeted for improvement of drought tolerance in maize (Edmeades 2013). Osmoprotection by osmotic adjustment and antioxidant scavenging defense system, plant growth regulators, water channel proteins, stress responsive proteins, transcription factors and signaling pathways actively participate in conferring drought tolerance in crop plants.

3.3.1 Osmotic Adjustment

Osmotic adjustment is described as development of water gradient to increase the water influx for maintaining turgor by lowering osmotic potential. Osmotic adjustment help to maintain the tissue water status. Damaging effects of drought are minimized by accumulation of solutes in cellular cytoplasm and vacuole. Protection is provided by maintaining the turgor potential and physiological processes with the help of osmotic adjustment (Taiz and Zeiger 2006). Plant water status is determined by water potential, osmotic potential, turgor potential and relative water contents (Kiani et al. 2007). Relative water contents act as integrative index for estimation of drought tolerance. Stomata are closed followed by reduction in CO_2 accumulation which is the result of reduction in relative water contents under drought stress (Gindaba et al. 2004). Translocation of photosynthetic assimilates to the developing kernels is also maintained by sustainable regulation of photosynthetic rate and turgor potential (Subbarao et al. 2000). Osmoprotectants are categorized into two main groups; first group is comprised of nitrogenous compounds like, proline, polyols, polyamines and glycinebetaine whereas, second group consists of hydroxy compounds like polyhydric alcohols, sucrose, and oligosaccharides (McCue and Hanson 1990). Accumulation of organic solutes (proline, sugar alcohols, glycinebetaine and soluble sugars) and inorganic ions (K^+, Na^+, Ca^{2+}, Mg^{2+}, Cl^-, NO^{3-}, SO^{4-}, and HPO_4) are involved in osmotic adjustment (Morgan et al. 1986; Bajji et al. 2001; Shao et al. 2006; Blum 2011). Osmolytes or osmoprotectants are neutral, organic and non-toxic compounds for plants. These osmolytes protect the cellular proteins and cellular membranes against the dehydrating effects of drought stress (Yancey et al. 1994).

Glycinebetaine is organic, water soluble and non-toxic for plants and has very important role in protection of plants against drought, salinity, cold and heat stresses by acting as osmoprotectant (Ashraf and Foolad 2007; Chen and Murata 2008). Glycinebetaine protects photosynthetic apparatus, stabilize cellular proteins (RuBisCo), reduces load of reactive oxygen species and acts as osmoprotectant (Makela et al. 2000; Allakhverdiev et al. 2003). There are significant genetic differences among maize genotypes for accumulation of glycinebetaine under drought stress. Transgenic studies of maize using betA gene of *Escherichia coli* which encodes for choline dehydrogenase (enzymes involved in the pathway of glycinebetaine biosynthesis) showed improvement in drought tolerance. This improvement in drought tolerance was reported right from germination to economic yield in case of transgenic maize (Quan et al. 2004). Glycinebetaine confer drought tolerance in transgenic maize by maintaining cell membrane integrity and cellular enzyme activity. Exogenous application of glycinebetaine improves the kernel morphology, kernel quality (Ali and Ashraf 2011), leaf area, 100 kernels weight, biological yield, grain yield, harvest index, relative leaf water contents, proline contents, total soluble proteins and antioxidative defense (Anjum et al. 2012) of maize under water deficit conditions which proves that glycinebetaine is important osmolyte for drought tolerance.

Proline an imino acid, acts as osmoprotectant under different abiotic stresses like, drought salinity and extremes of temperature. Proline maintains water status by maintaining osmotic potential, protects cellular membranes, prevents denaturation of cellular proteins and maintains sub-cellular structures under osmotic stress (Ashraf and Foolad 2007). Proline is well known as osmo-protector, osmo-regulator and regulator of cellular redox potential (Mohammadkhani and Heidari 2008). Accumulation of proline contents is directly associated with drought tolerance. Proline being osmoprotectant, reported to be accumulated in maize under drought stress and conferred drought tolerance (Bänziger et al. 2000). Proline oxidation pathway is suppressed by down-regulation of proline dehydrogenase (PDH) enzyme (first enzyme for proline oxidation) in maize under drought stress. Down-regulation of PDH is ABA independent regulation which showed that ABA is not contributor for its down-regulation under drought stress (Bruce et al. 2002). Hundred times increase of proline contents in the primary roots of maize was reported under drought stress. Accumulation of proline is dependent on the coordinated mechanism of biosynthesis, catabolism and transportation within the plant under drought stress (Mohammadkhani and Heidari 2008).

Accumulation of soluble sugars increases in roots and shoots of maize under drought stress. This accumulation is dependent on degradation of starch. Ratio of soluble sugars to starch increases due to increase in accumulation of soluble sugars and decline in accumulation of starch because of starch degradation. Soluble sugars being hydrolytic products are very important for plant metabolism and act as substrate for biosynthesis processes, sugar sensing, signaling pathways, metabolic regulation and energy production. Soluble sugars protect the plants under drought stress either through substitution of water with hydroxyl group to maintain the hydrophilic interaction with proteins and membranes or secondly through vitrification. Vitrification is involved in the formation of biological glass in cytoplasm to protect the cellular organelles under drought stress. Drought tolerance is reported to be positively associated with accumulation of soluble sugars under drought stress (Mohammadkhani and Heidari 2008).

Polyols/sugar alcohols well-known osmoprotectants, are reduced aldose and ketose sugars. Cyclic polyols are ononitol and pinitol whereas; acyclic polyols are sorbitol, glycerol and mannitol (Abebe et al. 2003). Hydroxyl group of polyols make hydration sphere around the macromolecules and protect them from dehydration (Williamson et al. 2002). Sorbitol increases the accumulation of proline, nitrate reductase (involve in nitrate assimilation) and nitric oxide (NO) in maize under osmotic stress (Jain et al. 2010).

However, it is reported that osmotic adjustment (OA) is governed by dominant gene effects rather than additive gene effects and genetic variability for OA is low among tropical maize populations which shows that small genetic gain can be achieved by selection. Furthermore, genetic variability among tropical maize germplasm for OA is narrow at vegetative stage relative to reproductive growth stages which recommends that selection based on OA should be made during reproductive growth stages. Weak correlation between yield and OA is obvious in maize under drought stress which proves that improved OA brings very little betterment in kernel yield. These results are specific to two tropical maize populations and furthermore genetic variability might be present across the regions and across the seasons (Guei and Wassom 1993).

3.3.2 Antioxidative Defense Mechanism

Specific molecules which prevent the oxidation of other molecules by scavenging reactive oxygen species are known as antioxidants. Antioxidants (AOX) act as defense shield for plants against oxidative damage caused by reactive oxygen species. Enzymatic and non-enzymatic compounds are part of antioxidant defense system. Enzymatic components comprised of catalase (CAT), superoxide dismutase (SOD), glutathione reductase (GR), ascorbate peroxidase (APX), peroxidase and polyphenol oxidase while non-enzymatic antioxidant compounds are α-tocopherol, ascorbic acid, β-carotene, glutathione and cysteine (Gong et al. 2005). High contents of enzymatic and non-enzymatic antioxidants are critically important for conferring drought tolerance in plants. Sequential generation of reactive oxygen species (ROS) in response of osmotic stress has been depicted in Fig. 3.2. Hydrogen peroxide (H_2O_2) and superoxide radicals are scavenged by ascorbate-glutathione cycle which consists of peroxidases, catalases, peroxiredoxins, ascorbate peroxidase, monodehydroascorbate reductase, dehydroascorbate reductase and glutathione reductase (Fazeli et al. 2007). Sequential scavenging of ROS by AOX has been described in Fig. 3.3.

Enzymes of ascorbate-glutathione cycle are localized in stroma of chloroplast, cytosol, mitochondria and peroxisomes. Superoxide molecules are dismutated into O_2 and H_2O_2 by superoxide dismutase in first step of ROS scavenging. Carotenoids have potential capability of scavenging lipid peroxy-radicals and singlet oxygen. Carotenoids prevent the generation of superoxides and lipid peroxidation under drought deficit conditions (reviewed by Farooq et al. 2009). Hydrogen peroxide is

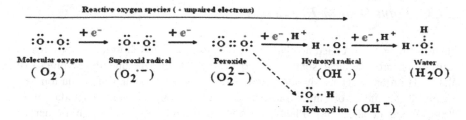

Fig. 3.2 Sequential generation of reactive oxygen species in response of drought stress (Cristiana et al. 2012). © 2012 C Filip, N Zamosteanu, E Albu. Originally published in [shortcitation] under CC BY 3.0 license. Available from: http://dx.doi.org/10.5772/47795

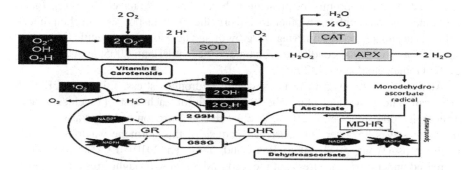

Fig. 3.3 Sequential scavenging of reactive oxygen species by antioxidant defense system (Ceron-Garcia et al. 2012). © 2012 A Ceron-Garcia, I Vargas-Arispuro, E Aispuro-Hernandez, MA Martinez-Tellez. Originally published in [short citation] under CC BY 3.0 license. Available from: http://dx.doi.org/10.5772/26057

detoxified by peroxidase and catalase (Apel and Hirt 2004). Dismutation by superoxide dismutase confers frontline defense shield against superoxide radicals so, contents of superoxide dismutase are directly linked with oxidative stress tolerance (reviewed by Farooq et al. 2009). Comparison of well-known drought tolerant and susceptible maize genotypes showed that APX, GR, CAT and POX were high in value in tolerant genotypes with increase of drought stress at reproductive stages but these contents were decreased when drought prevailed for extended period. In case of susceptible maize genotypes, these AOX components were reduced with initiation of drought stress. H_2O_2 and malondialdehyde (MDA) contents of tolerant genotypes were lower even under drought stress which showed that higher AOX contents have quenched the ROS and prevented the cellular membrane damage by showing lower MDA contents (Chugh et al. 2013). Another comparison among maize genotypes was conducted at seedling stage, which showed that antioxidants were higher in tolerant genotypes and lower in susceptible genotypes. Lipid peroxidation of cellular membranes and ROS load was lower in drought tolerant genotypes due to strong AOX defense system (Chugh et al. 2011).

3.3.3 Plant Growth Regulators

Plant hormones also called as plant growth regulators, phytohormones and growth substances, are chemical substances governing the growth and development of plants. Hormones act as signaling molecules, trigger cellular differentiation, act locally at the site of origin or transported to distant targets (Gomez-Roldan et al. 2008; Wang et al. 2008). Plants respond to drought stress through numerous adaptations and one of those is endogenous hormonal balance. Different plant growth regulators confer drought tolerance in plants. Auxins, Cytokinins (CKs), Abscisic acid (ABA), Indole acetic acid (IAA), Gibberellin3 (GA3), Salicylic acid (SA), Brassinosteroids (BR), Methyl jasmonate (MeJA), Polyamines, Ethylene and Zeatin (ZT) are the prominent plant growth regulators (Wang et al. 2008). Plant hormones interact with each other to govern the plant responses and their interaction is dependent on specific stage of growth and development, specific tissue and specific environmental conditions (Emam and Seghatoeslami 2005; Weiss and Ori 2007; Gomez-Roldan et al. 2008).

Auxins were involved in responses against drought stress. Ethylene, CK and auxin are interactive and effect the biosynthesis of each other (Tsuchisaka and Theologis 2004; Jones et al. 2010). Decline in concentration of IAA was observed in maize leaves however, inconsistent response of this hormone was observed in other crops (Wang et al. 2008; Bano and Yasmeen 2010). IAA accumulation increased under moderate stress but reduced under severe drought stress. Decline in IAA was due to increased degradation, reduced biosynthesis and higher ABA accumulation (Pirasteh-Anosheh et al. 2013). IAA accumulation increased 13.4 % under moderate stress whereas; decreased 63.2 % under severe drought stress in maize.

GA_3 and ZT reduces in maize leaves when plants are subjected to drought stress (Wang et al. 2008). Decline in GA_3 accumulation are either due to reduced biosynthesis or increased degradation of this hormone by ROS (Pirasteh-Anosheh et al. 2013). Increased ABA accumulation is one of the reasons for decline in GA_3 accumulation and 68 % reduction in GA_3 accumulation was observed (Wang et al. 2008). ZT is most sensitive phytohormone which is reduced up to 73.3 % in maize leaves under drought stress (Pirasteh-Anosheh et al. 2013).

Salicylic acid (SA) is involved in plant growth, flower induction, thermogenesis, nutrient uptake, stomatal regulation, ethylene biosynthesis, enzyme regulation and photosynthesis (Hayat & Ahmed 2007). Photosynthesis is maintained by SA through retaining the higher chlorophyll contents under drought stress and confers drought tolerance (Rao et al. 2012).

Plant hormonal balance acts as regulator for different processes of growth and development. Environmental factors and plant growth stages are determinants of hormonal balance. ABA and ethylene govern stomatal conductance, number of grains, grain filling rate and growth of plant apex (root and shoot) in antagonistic way. Effects of ABA as stomata closing agent are suppressed by combined higher accumulation of ethylene and CKs. CKs enhance the growth, development and yield by improving the stay-green index, qualitative and quantitative characteristics of grain. Effectiveness of these hormones depends on interaction between them (Wilkinson et al. 2012).

ABA as a stress hormone is actively involved in modulation of growth, development and responses against stresses. ABA dependent signaling pathway comprised of numerous components like light-harvesting chlorophyll a/b binding proteins, chloroplast envelope-localized ABA receptor, chloroplastic antioxidant defense enzymes and these components are highly responsive to ABA (Shen et al. 2006; Hu et al. 2007, 2008). Plastids in plants have thousands of proteins. Chlorophyll biosynthesis and expression regulation of chloroplast proteins is dependent on drought stress (Lawlor and Cornic 2002; Leister 2003). Maize mutants were developed with deficiency in ABA biosynthesizing enzymes for assessment of protein regulation in leaves. Study revealed that expression of chloroplast proteins was dependent on ABA under drought and light prevalence as ABA deficiency made the leaf devoid of photosynthesis and antioxidants. Antioxidant defense mechanism and photosynthesis are found to be ABA dependent in maize mutants which shows that maintenance of photosynthesis and antioxidant defense are critical adaptation for drought tolerance (Hu et al. 2012). Normal water availability does not induce even minute amount of ABA and extremely severe drought reduces the ABA accumulation due to cessation of ABA precursors like carotenoids and xanthophylls (Pirasteh-Anosheh et al. 2013).

Plants responses are modulated by polyamines against drought, oxidative stress, heavy metal toxicity, salinity, osmotic stress, temperature extremes and flooding. Polyamines act as signals and messengers to regulate the plant responses for development of tolerance against stresses (Gill and Tuteja 2010). Polyamines are organic compounds of low molecular weight with diverse functions due to having diverse position and numbers of amino acids. These cationic ions have crucial role in stabilization and destabilization of cellular contents through electrostatic linkage with DNA, RNA and proteins. Polyamines involved in numerous developmental processes, survival of plant embryos, membrane stabilization, cell signaling, modulation in gene expression, cell proliferation, apoptosis, cell death, morphogenesis, cell differentiation, germination, seed dormancy, flower induction, embryogenesis, grain ripening and aging (Rea et al. 2004). Polyamines are present in the form of putrescine (Put), spermidine (Spd) and spermine (Spm). ABA accumulation is responsible for increase in accumulation of polyamines in maize (Liu et al. 2005). Photosynthetic pigments, leaf area and dry matter of maize increase by exogenous application of polyamines. Accumulation, absorption and compartmentalization of K^+, Mg^{++} and Ca^{++} increase in tolerant maize genotypes relative to susceptible ones under drought stress. Potassium contents increase in roots and shoots of tolerant maize genotypes (Shaddad et al. 2011). Ca^{++} and Mg^{++} accumulation and absorption reduce in roots and shoots of maize under drought stress but these improve by exogenous application of polyamine and phytohormones. Potassium ions improve leaf water potential, relative water contents, turgor potential, transpiration rate, photosynthetic rate, grain weight per cob, economical yield and kernel yield (Aslam et al. 2014c) whereas, osmotic potential improved significantly across the years.

Polyamines are positively associated with kernel yield only under mild water deficiency whereas; under severe drought stress these have nothing to do with yield. Polyamines are more effective to incorporate drought tolerance relative to other phytohormones. Drought tolerant maize genotypes are more responsive to exogenous application of polyamines and phytohormones whereas, susceptible genotypes are least responsive (Shaddad et al. 2011). These small ionic molecules are reported to be involved in osmotic adjustment under drought stress in different crops (Shao et al. 2006; Liu and Bush 2006).

Brassinosteroid (BR) is steroid plant hormone which governs growth, development, numerous physiological processes, and tolerance against biotic and abiotic stresses (Bajguz 2007; Ren et al. 2009; Tanaka et al. 2009; Xia et al. 2009). Improvement of drought tolerance in maize by exogenous application of BR shows that this hormone is involved in important plant responses (Li et al. 1998). BR alleviates the oxidative damage caused by reactive oxygen species in maize under osmotic stress. Endogenous and exogenous BR availability is significantly and effectively responsible for conferring tolerance against oxidative damage in maize. BR-induced osmotic stress tolerance was dependent on ABA whereas, ABA-induced osmotic stress tolerance is not dependent on BR and their molecular background is elusive. BR upregulates the vp14 gene in maize leaves which is involved in biosynthesis of ABA. BR increases the nitric acid in mesophyll cells of maize leaf which regulates the ABA biosynthesis and confers the oxidative stress tolerance (Zhang et al. 2011).

Methyl jasmonate (MeJA) is naturally occurring plant growth regulator and acts as modulator of numerous morphological, physiological and biochemical plant processes. MeJA actively participates in photosynthesis, plant growth, cell division and stomatal closure (Ueda and Saniewski 2006; Norastehnia et al. 2007; Anjum et al. 2011a) and its role in incorporation of drought tolerance across plant species is very well known. Methyl-jasmonate confers drought tolerance in maize by upregulating the osmoprotectants (proline, free amino acids and soluble sugars), antioxidative defense and phytohormones (Abdelgawad et al. 2014).

In hormonal balance each hormone does not work independently but there is strong interaction between them for regulation of growth and developmental processes under drought stress. ABA has antagonistic effects on other hormones like Auxin, IAA, GA_3 and ZT. ABA acts as growth suppressor and others (Auxin, IAA, GA_3 and ZT) act as growth promoter so, their ratio provides the evidence for growth responses (Weiss and Ori 2007).

3.3.4 Molecular Mechanisms of Drought Tolerance

Molecular events of the cell are affected in response of abiotic stresses. Water channel proteins, stress responsive proteins, transcription factors and signaling pathways are the molecular events which are responsive to drought stress. These molecules are involved in conferring drought tolerance through protection of cellular contents or by regulation of stress responsive genes.

3.3.4.1 Stress Proteins and Water Channel Proteins

Certain proteins have critical role in imparting tolerance against drought stress, broadly these are known as stress proteins. Stress proteins are mostly water soluble and ensure tolerance by hydration of cellular contents (Wahid et al. 2007). Late embryogenesis abundant (LEA)/dehydrin, heat shock proteins (HSP), cold shock proteins, aquaporins and Cyclophilins (CYP) are certain proteins which have critical role in enhancing drought tolerance. Aquaporins are water channel proteins and confer drought tolerance by increasing water uptake through provision of opened water channel.

LEA is a group of proteins which naturally accumulate in pollen grains, seeds and vegetative tissues during prevalence of abiotic stresses. Accumulation of LEA proteins is directly associated with desiccation tolerance in crop plants (Amara et al. 2012). LEA proteins are categorized into numerous classical groups based on amino acid sequences, specific domains, distinct motifs and peptide composition (Battaglia et al. 2008). Among these different groups, 1, 2 and 3 are main groups which consist of large number of LEA proteins. Proteins belonging to Groups 1 undergo complex post-translational modifications (PTMs) including, acetylation, phosphorylation, deamination and methylation and these PTMs depict their seed specific role. Anti-aggregative properties are characteristic features of group-2 and group-3 LEA proteins. Groups 2 LEA proteins provide protection to membranes by preventing their denaturation. Groups 3 LEA proteins are linked in dehydration tolerance and prevent the cell shrinkage which is the result of water loss. EMB564, MLG3 and RAB17, are the representative LEA proteins of group1, 2, and 3 respectively for assessment of functional characteristics in maize (Amara et al. 2012).

Heat shock and cold shock proteins are chaperones which prevent the denaturation and folding of cellular contents under stresses. Small heat shock proteins (sHSP) constitute the larger HSP groups which are the indication of tolerance against drought stress and heat stress. Cold shocks proteins (CSPs) accumulate under stress and are the chaperones for protecting dysfunction of RNA. CSPA and CSPB genes from bacteria were transgened in maize which increased the tolerance against drought, heat and cold stresses by protecting photosynthesis, vegetative and reproductive growth stages. Yield increased (11–21 %) in CSPA and CSPB transgenic maize as evaluated through multi-year, multi-location and multi-hybrid yield trials (Yang et al. 2010). Qualitative differences in the biosynthesis of heat shock proteins were observed in maize lines. There are differences present among maize lines for production of total proteins, low molecular and high molecular weight heat shock proteins (Ristic et al. 1991). sHSP17.2, sHSP17.4 and sHSP26 are three identified small heat shock proteins by using MALDI-TOF mass spectrometry in maize leaves in response to drought stress. Transcriptional analysis showed that expression level of these sHSP is upregulated in response of heat and drought stress in maize leaves. ABA was found to be responsible for post-transcriptional expression regulation of these sHSPs (Hu et al. 2010).

Cyclophilin (CYP) proteins are involved in multiple functions like, cell division, cell signaling, protein trafficking, transcriptional regulation, pre-mRNA splicing and stress tolerance (Trivedi et al. 2012). Silico analysis revealed that stress response was positively associated with cyclophilin proteins. CYP proteins are more than 80 % similar in rice, maize, arabidopsis, sorghum and brachypodium which show their high frequency of conserved sequences. CYP is frequently found in chloroplast, cytosol, lumen and mitochondria (Trivedi et al. 2013). Comparative evaluation showed that mRNA responsible for the synthesis of CYP was found in high frequency in maize after 6–7 h of stress treatment whereas; in beans it took about 48 h for that level. Gene family of 6–7 members encodes for CYP in maize but only single gene is responsible in bean (Marivet et al. 1992).

Aquaporins act as facilitator for osmosis through water channels and increase the membrane permeability. Expression of aquaporins is differentially effected in maize under drought stress. Accumulation of large number of aquaporins in tolerant maize genotypes shows that these have critical role in conferring the drought tolerance. Tonoplast intrinsic proteins (TIP), nodulin-like intrinsic proteins (NIP), and plasma membrane intrinsic protein (PIP), subfamilies of aquaporin proteins are upregulated in maize (Hayano-Kanashiro et al. 2009).

Stress responsive proteins are very effectively involved in conferring drought tolerance in maize genotypes. Selection of maize genotypes based on these stress responsive proteins and their use in different hybridization program can further improve drought tolerance potential. Aquaporins provide water channel and increase rate of permeation so, genotypic selection based on higher contents of aquaporins proteins under stress conditions can further increase water uptake and improve drought tolerance.

3.3.4.2 Transcription Factors

Natural master regulators of cellular processes and modifier of traits in response of stress are transcription factors. Single gene engineering which encodes for specific protein is not sufficient to confer tolerance because of complex tolerance mechanism (Bohnert et al. 1995). Transcription factors (TFs) have the potential to regulate multistep complex pathways by modifying the metabolite fluxes in predictable pattern. TF families in plants are extensively large compared to animals and microorganisms. Stress responsive pathways are regulated at the level of transcription factors. Studies of transcriptional factor regulation were mostly conducted on model plant *Arabidopsis thaliana*. TFs of more than 50 different families encoded by 1700 different genes were reported in Arabidopsis (Yang et al. 2010). In present study we are going to discuss only those TFs which are responsive to drought stress and confer drought tolerance. AP2/EREBP [DRE binding protein

(DREB)/CRT binding factor (CBF)], basic leucine zipper (bZIP) like ABA responsive element binding protein (AREB)/ABRE binding factor (ABF), zinc-finger e.g. C2H2 zinc finger protein (ZFP), NAM (no apical meristem), CUC2 (cupshaped cotyledon), ATAF1-2, CCAAT-binding e.g. nuclear factor Y (NF-Y) and NAC e.g. stress-responsive NAC (SNAC) are known to have their crucial role for drought tolerance (Shinozaki et al. 2003; reviewed by Yang et al. 2010).

ABA responsive element binding factors (AREB/ABF) are member of basic leucine zipper (bZIP) TF family; they are involved in ABA signaling under drought stress and seed maturation. AREB/ABF recognizes and binds ABA responsive element (ABRE) and conserves cis-element to regulate the expression of downstream genes (Mundy et al. 1990). ABRE is located in the promoter region of ABA responsive genes (Yamaguchi-Shinozaki and Shinozaki 2006). Stomatal closure and reduced transpirational losses are the result of increased ABA contents due to over expression of AREB1/ABF2, ABF3 or AREB2/ABF4 (Kang et al. 2002). Over expression of ABF3 confer drought tolerance through increased chlorophyll fluorescence (Fv/Fm) and reduced leaf wilting and rolling (Oh et al. 2005).

NAC (NAM, ATAF1-2, CUC2) is family of TFs which are specific to plants; it consists of highly conserved DNA binding NAC domains (Guo et al. 2008). NAC019, NAC055 and NAC072 are members of NAC TF family, they recognize and bind to NACRS (NAC recognition sequence, CATGTG) element which is located in promoter region of early responsive to dehydration (ERD1) and confer tolerance against dehydration (Tran et al. 2004). About 40 NAC TFs out of 140 are found to be responsible for drought and salinity tolerance in rice. Over expression of SNAC1 reduces the stomatal aperture in response to drought stress. SNAC1 in transgenic plants increases 17–22 % spikelet fertility and 22–34 % seed setting under moderate and severe drought stress, however under well watered conditions there were differences between transgenic and non-transgenic plants (Hu et al. 2006). TF is governed by maize ubiquitin (ZmUbi) promoter (Nakashima et al. 2007). OsNAC6 rice transgenic seedlings increases 42–57 % recovery after removal from hydroponic growth medium.

Certain TFs are responsive to dehydration but not to ABA, these are called ABA-independent dehydration-responsive TFs and these are different from AREB/ABF and SNAC which are ABA responsive. Conserved Cis-element localized in promoter region of concerned gene is potential binding site for these TFs, which is known as dehydration responsive element (DRE). This typically ABA-independent dehydration responsive TF family is known as DREB TF family. These TFs act as response regulator under drought stress, cold stress and for the developments of leaf, flower and seeds (reviewed by Yang et al. 2010). Two types of DREB (DREB1 and DREB2) are found responsive for different stresses. DREB1 is responsive to cold stress whereas, DREB2 is responsive to drought, heat and salt stresses. For improvement of stress tolerance in crops DREB1 signaling pathway is extensively explored. Constitutive over-expression of ZmDREB1A (DREB1A of *Zea mays* L.) and OsDREB1A (DREB1A of *Oryza sativa* L.) in arabidopsis upregulated the downstream genes and conferred tolerance against drought, heat and salinity stresses (Qin et al. 2004; reviewed by Yang et al. 2010). Transgenic arabidopsis for DREB1A genes of maize and rice followed by improved stress shows that DREB1 function and pathway is highly conserved in monocot and dicot species. ZmUbi (*Zea mays* L. Ubiquitin) promoter is effectively regulat the DREB1A, DREB1B and DREB1C genes in arabidopsis for improvement of tolerance against drought stress (reviewed by Yang et al. 2010).

HARDY is another member of DREB subfamily which is expressed in inflorescence tissues. The constitutive over expression of this member has pleotropic effects on vegetative growth and forms dense roots and thick green leaves. HARDY improves water use efficiency through improving photosynthesis and reducing transpiration (Karaba et al. 2007). DREB2A has little role for conferring drought tolerance in arabidopsis. Over expression of DREB2 TFS like ZmDREB2A (maize) and GmDREB2 (soybean) improved drought, heat and salinity tolerance. Generally DREB2A needs certain modifications in Arabidopsis for improvement of drought tolerance whereas, DREB2A of maize and soybean confer drought tolerance without any modification (Chen et al. 2007; Qin et al. 2007).

Zinc Finger Protein (ZFP) TFs, are strongly induced by salinity, drought, ABA and cold treatments which depicts that these TFs are dynamically involved in stress responses through functioning as transcriptional repressors or activators (Sakamoto et al. 2000, 2004). Members of ZFP TF family like Drought and Salt Tolerance (DST; Huang et al. 2009), OsZFP252 (Xu et al. 2008) and ZAT10 (Xiao et al. 2009) are actively involved in regulation of dehydration responses. ZFP252 is found to be located upstream of DREB1A TF and upregulated the accumulation of soluble sugars and proline. DST, a negative regulator, improves the salt and drought stress tolerance without harboring yield losses through its knockdown expression (Huang et al. 2009). C2H2 ZFP, EAR (ERF-associated amphiphilic repression), cytoplasmic (non-TF) ZFP, and ZAT10 are the members of ZFP TF and improves drought tolerance in rice (reviewed by Yang et al. 2010).

Nuclear Factor (NF-Y) consists of A, B and C subunits and belongs to CCAAT-binding transcription factor family. AtNF-YB1 and AtNF-YA5 improves drought tolerance in Arabidopsis. ABA and drought stress are responsible for induction of AtNF-YA5 which resultantly close stomata by reducing aperture (Li et al. 2008). Constitutive overexpression of ZmNF-YB2 in maize improves drought tolerance which is accomplished through higher photosynthesis rate, low leaf rolling, high stomatal conductance and low leaf temperature. Under drought stress there was 50 % increase of yield whereas, under normal conditions there were slightly compressed internodes and earliness of flowering in ZmNF-YB2 transgenic relative to non-transgenic maize (Nelson et al. 2007).

ASR1 is a putative transcription factor which encodes for stress responsive protein and ABA. Differential expression of ASR1 is responsible for determination of leaf size and leaf senescence. Over-expression induces shorter leaves whereas reduced expression triggers larger leaves in ASR1 transgene. ASR proteins protect the DNA structure through maintaining the DNA topology in response of deficit water. Re-channelization of assimilates from source to sink followed by senescence of source is also one of the ASR1 functions (Jeanneau et al. 2002).

There is huge variability reported among TFs across the plant species. But there is still dire need to exploit their potential at best. Most of the literature shows that role of TFs is assessed under screenhouse conditions and seedling stages. In screenhouse conditions, drought stress is imposed for few hours whereas, in field conditions duration of drought stress can prolong for many days. Assessment of tolerance by TFs at seedling stage is also not true representative of yield. So, for assessment of real time functional characterization of TFs, evaluation must be made at reproductive growth stages under field conditions which will be true contribution in economical yield through drought tolerance.

3.3.4.3 Signal Transduction Pathways

Adaptive responses in plants are broadly categorized into three categorical groups, (a) osmotic adjustment or osmotic homeostasis; (b) stress damage control, detoxification and repair, (c) growth control (reviewed by Zhu 2002). Depending on plant responses, drought stress signaling is again categorized into three functional characteristics groups; (a) osmotic stress signaling for restoration of cellular homeostasis; (b) detoxification stress signaling to prevent cell damage and to repair cell damages; (c) signaling to maintain cell division and cell expansion to sustain growth. Homeostasis and detoxification signaling confer drought tolerance and regulate the stress responses to maintain growth (reviewed by Zhu 2002).

Osmotic stress signaling is accompanied by protein phosphorylation. Protein kinases are activated in response of osmotic stress. Calcium signaling in response of osmotic stress stimulates calcium-dependent protein kinases (CDPK) which further regulates downstream responses. Constitutive overexpression of CDPK in maize protoplast regulates the expression of certain genes which are responsive to ABA, cold and osmotic stresses (Sheen 1996). These findings link the induction of gene expression in response of osmotic stress with calcium signaling. Transcriptome for protein kinases like, MAPK, MAPKK, MAPKKK and histidine kinase is increased in response of osmotic stress (Mizoguchi et al. 2000). Understanding of signaling pathways depends on clear insight of inputs and outputs of pathway. Osmotic stress, changes in turgor and subsequent injury are inputs whereas, osmotic adjustment by osmolytes, protection from damage and repair from damage (induction of LEA or dehydrins) are outputs of osmotic signaling pathway (reviewed by Zhu 2002).

Membrane phospholipids (dynamic mechanism) produce numerous signaling molecules (IP3, DAG, PA) and provide structural support to cells in response of osmotic stress. At low level of osmotic stress, phospholipids act as messengers and regulate downstream responsive genes whereas, under severe osmotic stress, high level of phospholipids depict the cellular damage (Sang et al. 2001). Phospholipases and other messengers are the basis for categorical grouping of phospholipid signaling systems. Inositol 1, 4, 5-trisphosphate (IP3), a secondary messenger induces stomatal closure by increasing Ca^{+2} ions in cytoplasm. Calcium ion accumulation in cytoplasm is responsible for regulation of osmotic stress responsive genes as reported through pharmacological and microinjection experiments (Wu et al. 1997).

IP3 is found to be osmotic stress responsive through numerous experiments which shows that overexpression of ABA and stress responsive genes are directly associated with higher IP3 accumulation (Sanchez and Chua 2001). Other inositol phosphates, like, IP_6, I_1P_3, I_3P_3 and I_4P_3 are involved in release of calcium ions from internal cellular sources (reviewed by Zhu 2002). Phospholipase C (PLC) and phospholipase D (PLD) are important components of osmotic stress responsive signaling pathway. Phosphatidic acid (PA) is a secondary messenger and is produced by PLD through cleavage of membrane phospholipids. PLD enhances lipolitic membrane disintegration and its activity is found to be higher in susceptible genotypes. ABA treatment increases the PLD activity whereas; PA treatment reduces the ABA effects (Jacob et al. 1999).

Osmotic stress tolerance and plant water balance regulation are among the main characteristic functions of ABA. ABA mutants are developed in numerous crops including maize and these mutants are unable to grow under drought and temperature stresses (Koornneef et al. 1998). ABA mutants produce short stature plants which show the involvement of ABA in regulation of cell cycle and other cellular activities. Molecular mechanisms lying behind the higher accumulation of ABA in response of osmotic stress became clear through recent advances in molecular biology. Accumulation of ABA depends upon the equilibrium of ABA biosynthesis and degradation. It is reported that numerous ABA biosynthesizing genes like ZEP, 9-cis-epoxycarotenoid dioxygenase (NCED), ABA aldehyde oxidase (AAO) and LOS5/ABA3 are upregulated in response of drought stress (reviewed by Zhu 2002). Genes involved in degradation of ABA like cytochrome P450 monooxygenase, are downregulated in response of drought stress (reviewed by Zhu 2002).

ABA dependent gene upregulation is initiated after increase in ABA biosynthesis in response of drought stress. Genes involved in osmotic homeostasis like, aquaporins, osmolyte, LEA, dehydrins, chaperones, detoxification enzymes, ubiquitination associated enzymes and proteases are upregulated (Zhu et al. 1996; Zhu 2002). Drought responsive genes are broadly classified into two groups; early response and delay response genes. Induction of early response genes starts very quickly even within minutes and encodes for transcription factors. Activated TFs control further subsequent delay gene responses. Delay response genes are induced after few hours of stress imposition and comprise of large number of stress responsive genes. CBF/DREB gene family, AtMyb, ABF or ABI5 or AREB and RD22BP are examples of early response genes for ABA regulation under drought stress, salt stress, and cold stress (reviewed by Zhu 2002).

ABA intensive and ABA deficit mutants are developed to determine ABA dependent responses. ABA intensives are not completely intensive and ABA deficits are not completely deficit; certain responses are independent of ABA. Numerous studies reported that some genes are totally ABA dependent, some are totally ABA independent and some are partially dependent (Shinozaki and Yamaguchi-Shinozaki 1997). RD29A is an excellent example of ABA dependent and ABA independent signaling. DRE (dehydration responsive element) sequence is identified to be located in *RD29A* gene and activation of this element is independent of ABA (Xiong et al. 2001).

Chapter 4
Global Achievements in Drought Tolerance of Maize

Contribution of public sector in improvement of drought tolerance is meaningful as they developed methodologies and genetic resources. Comparison of public sector with multinational private sector revealed that public sector is lacking in real professional and technical staff, sustainable resources, discipline, and coordination to develop the drought-tolerant material on consistent basis. Public sector projects are dependent on continuous availability of funds. Private sector has done tremendous work for improvement of maize especially against drought stress. Pioneer and DeKalb hybrids (private companies) were credited for development of drought-tolerant maize hybrid for temperate zone (Castleberry et al. 1984). From 1950 to 2001, 18 elite commercial hybrids were released for cultivation (Duvick et al. 2004). Experiments showed that over the time selection has brought about numerous changes in behavior and response of germplasm. As, leaf rolling scores were higher in 1950s and reported to be lowered in 2000, whereas higher rolling scores means the flatness or erectness of leaves and lower scores showed the increased rolling. Tassel branch number and variation in stem volume were also reduced from 1950 to 2000 due to extensive selections (Edmeades et al. 2006).

Improvement in drought tolerance is the output of extensive multi-environment trials. Drought stress prevailed during flowering stage is purposed to be most critical in effecting the kernel yield in maize. Different hybrids were compared which belongs from different chronological time periods. Maize hybrids released in 1940s, when subjected to drought stress at flowering, these hybrids yielded 2.20 t/ha, whereas maize hybrids which were released in 1990s yielded 7.19 t/ha. These hybrids were also subjected to drought stress at grain filling, and their yield was 4.97 and 8.69 t/ha, respectively (Barker et al. 2005). These results showed that improvement in tolerance and increased genetic gain under drought stress were due to increased kernel setting or reduced barrenness. Whereas Hammer et al. (2009) attributed the genetic gains in yield to increased water uptake in modern hybrids due to extensive root system. From 1940s to 1990s, irrigation water had increased the grain production up to threefolds due to increased water uptake and increased yield potential and drought tolerance (Butzen and Schussler 2009). Continuous evidences have supported that increase in grain yield and drought tolerance in modern U.S. Corn Belt germplasm are associated with root volume and root intensity (http://www.asgrowanddekalb.com/products/corn/Pages/rootdig.aspx).

© The Author(s) 2015

M. Aslam et al., *Drought Stress in Maize (Zea mays L.)*,
SpringerBriefs in Agriculture, DOI 10.1007/978-3-319-25442-5_4

4.1 Contribution of CIMMYT, IITA, and Other Collaborative Partners

Breeding work is being carried out in different countries across the world by local and international research institutes. Research experiments proved that betterment in resistance against drought stress caused no yield penalty relative to normal water availability. However, selection can be done for alteration in floral parts and reproductive efficacy by changing the biomass partitioning within and to the maize ear (Edmeades 2008). Pedigree maize breeding program was used for selection of maize for reproductive traits under drought stress to improve the drought resistance (Bänziger and Araus 2007). Along with grain yield, secondary traits were given due importance for selection which have greater heritability under drought stress. CIMMYT worked out at Zimbabwe and Kenya, which are key selection centers and initial selection showed greater genetic gains under drought stress. CIMMYT-selected hybrids were compared with commercial hybrids across 36–65 locations in Southern Africa, and results showed 13–20 % higher yield in 1–5 ton/ha yield range and 3–6 % increase in the range of 5–10 ton/ha yield (Bänziger et al. 2006). Preliminary improvement in CIMMYT trails, increased the interest of donor agencies so, The Water Efficient Maize for Africa (WEMA) and The Drought Tolerant Maize for Africa (DTMA) projects were funded for the duration of 10 years by Bill and Melinda Gates Foundation.

DTMA Project was initiated in 2007, with collaboration of CIMMYT, *International Institute of Tropical Agriculture* (IITA) and 13 national institutes from sub-Saharan Africa. DTMA focused on conventional selection methods and marker-assisted selection (MAS) to genetically improve the maize germplasm which was already adapted to drier sub-Saharan conditions. Conventional pedigree hybrid breeding and biparental marker-assisted recurrent selection (MARS) were focused techniques in DTMA. Regional location centers in Zimbabwe, Kenya, Nigeria, and Zambia were selected for phenotyping of maize germplasm. In DTMA project, association mapping was also accomplished to identify the genomic regions associated with drought and heat tolerance (either drought and heat alone or their combination) in 293 inbred lines (Cairns et al. 2013). Large-effect QTLs which contribute more than 10 % of phenotypic variance were not identified in maize for drought tolerance. So MARS was modified into genomic selection based on genome estimated breeding values (GEBVs). DTMA and WEMA projects also developed a distinct database comprised 5000 lines which belongs to 27 interrelated populations. This database is an excellent source for genetic studies on drought tolerance of maize in tropical germplasm (DTMA 2012). It was also revealed in DTMA project that some source lines of drought tolerance became susceptible with increase in temperature which highlighted that selection for tolerance against drought and heat stress should be carried out simultaneously (DTMA 2012).

Water Efficient Maize for Africa (WEMA) was initiated in 2009, with collaboration of CIMMYT, Monsanto and five other eastern and southern African countries. Conventional selection, marker-assisted recurrent selection (MARS) and transgenic maize development was key focus of WEMA. Maize researchers seems to be confident that combination of doubled haploid (DH) inbred lines, genome-wide association systems (GWAS) and precision field-based phenotyping can bring the two times increase in genetic improvement of yield and drought tolerance (Bernardo 2008; Lorenz et al. 2011; Yan et al. 2011).

Maize in Asia is also suffering from drought stress so, there were two projects for drought tolerance of maize which were funded by GCP or the Syngenta Foundation and led by CIMMYT for South East Asia. China, Indonesia, India, Philippines, Thailand, and Vietnam are the component countries for Asian projects. Asian Maize Drought Tolerance (AMDROUT) Project was governed by CIMMYT and IITA as main partner. This Project was based on marker-assisted recurrent selection (MARS) and genome-wide selection (GWS). Yellow drought-tolerant maize inbred lines were developed in this project. This project was direly needed because about 80 % of Asian maize is grown in rainfed conditions. Asia will further face more severe drought stress in 2020s, and there is dire need to adapt the cropping pattern to changing climate. Pakistan, China, and Indonesia can adapt to changing climate due to their suitable geographical area, whereas other countries need priority breeding preferences for future. Phenotyping, marker-assisted recurrent selection (MARS), genome-wide selection (GWS) were the key techniques followed in this project. One cycle of phenotypic selection followed by one cycle of only genotypic selection give 50–100 % more genetic gain relative to two cycles of only phenotypic selection. So it is proved that GWS is beneficial technique for improvement of breeding populations. Biparental crosses were used between CIMMYT-Asia lines and African drought-tolerant donors (Vivek 2013). Best donor lines for drought tolerance were assessed through six location evaluation (China, Philippines, India, Thailand, Indonesia, and Vietnam) and results showed that CML444 was the best donor line. However, CML440, CML538, CZL0719, CML505, and CZL00009 donor lines were also proved to be effective donor. VL1012767, CML470, VL108729, VL108733, VL1012764, and CML472 were selected to be used as recipient parents (CGIAR Generation Challenge Programme 2014).

Genetic resources for better traits are present in germplasm but their evaluation is problematic due to linkage drag with non-desirable traits. However, to identify the drought-tolerant genes in germplasm, a project was funded by Mexico Government to CIMMYT and this project was named as Seeds of Discovery. *International Institute of Tropical Agriculture (IITA)* and CIMMYT breeding programs have established well-adopted pedigree breeding system for development of hybrids but there is still demand for open-pollinated varieties (OPVs) among the farmers of sub-Saharan Africa. OPV seed distributed from farmer to farmer without loss of performance. For development of farmer participation in selection, adoption, and production, mother–baby trail system was used in eastern and southern Africa. Mother–baby trials resulted in development of drought-tolerant OPVs like, ZM523, ZM521, and ZM409 (Edmeades 2013). Drought-tolerant maize hybrids and OPVs developed by CIMMYT in collaboration of other partner organizations are listed in Table 4.1.

Table 4.1 Drought-tolerant maize hybrids and OPVs

Country	Drought-tolerant hybrids	Drought-tolerant OPVs	Reference
Angola	CZH03030, CZH0819	ZM623, ZM423, ZM523, ZM725, ZM309	CIMMYT, IITA, DTMA (http://dtma.cimmyt.org/index.php/varieties/dt-maize-varieties)
Benin	–	Ku Gnaayi, Mougnangui, Ya Kouro Goura Guinm, Orou Kpinteke, Djéma- Bossi, DT SR W C2	CIMMYT, IITA, DTMA (http://dtma.cimmyt.org/index.php/varieties/dt-maize-varieties)
Ethiopia	MH130, MH138Q, MH140, BH546, BH547	Melkasa5, Melkasa6Q, Melkasa7, Gibe-2, Melkasa-1Q	CIMMYT, IITA, DTMA (http://dtma.cimmyt.org/index.php/varieties/dt-maize-varieties)
Ghana	Etubi, Enii-Pibi, Aseda, Opeaburoo, Tintim	Abontem, Omankwa, Aburohemaa, Wang Taa, Bihilifa, Sanzal-Sima, Ewul-Boyu	CIMMYT, IITA, DTMA (http://dtma.cimmyt.org/index.php/varieties/dt-maize-varieties)
Kenya	KDH3	KSCDT01	CIMMYT, IITA, DTMA (http://dtma.cimmyt.org/index.php/varieties/dt-maize-varieties)
Malawi	SC719, PAN53, CAP9001, MH27, MH28, MH30, MH31, MH32, MH33, MH34, MH35, MH36, MH37, MH38	ZM309, ZM523	CIMMYT, IITA, DTMA (http://dtma.cimmyt.org/index.php/varieties/dt-maize-varieties)
Mali	Tieba, Mata, Sanu, TZE-Y DT STR C4 x TZEI 13	Jorobana, Brico, Diambal	CIMMYT, IITA, DTMA (http://dtma.cimmyt.org/index.php/varieties/dt-maize-varieties)
Mozambique	SHIuvukani, Olipa, Molocue, Pris 601, SP-1	ZM309, ZM523, Dimba, Gema	CIMMYT, IITA, DTMA (http://dtma.cimmyt.org/index.php/varieties/dt-maize-varieties)
Nigeria	Sammaz 22, Sammaz 23, Sammaz 24, Sammaz 25, Oba Super 7, Oba Super 9, Ifehybrid 5, Ifehybrid 6	Sammaz 15, Sammaz 18, Sammaz 19, Sammaz 20, Sammaz 26, Sammaz 27, Sammaz 28, Sammaz 29, Sammaz 32, Sammaz 33, Sammaz 34, Sammaz 35, Sammaz 38	CIMMYT, IITA, DTMA (http://dtma.cimmyt.org/index.php/varieties/dt-maize-varieties)

(continued)

Table 4.1 (continued)

Country	Drought-tolerant hybrids	Drought-tolerant OPVs	Reference
Tanzania	WH403, WH502, WH505, VumiliaH1, HB405, HB513, HB623, TZH 636, TZH 538, TZH 417, NATA H104, NATA H105	ZM623, TZM 523	CIMMYT, IITA, DTMA (http://dtma. cimmyt.org/index.php/varieties/dt-maize-varieties)
Uganda	Longe 9H, Longe 10H, Longe 11H, UH 5051, UH 5052, UH 5053, UH5354, UH5355	VPMAX	CIMMYT, IITA, DTMA (http://dtma. cimmyt.org/index.php/varieties/dt-maize-varieties)
Zambia	KAM602, SC721, CAP5901, SC727, ZMS606, ZMS623, GV 635, GV 638, GV 628	ZM423, ZM523, ZM625, ZM721, Nelson's Choice	CIMMYT, IITA, DTMA (http://dtma. cimmyt.org/index.php/varieties/dt-maize-varieties)
Zimbabwe	ZAP51, ZAP61, Pris 601, ZS263, ZS265, PAN3 M-41, PGS53	ZM309, ZM401	CIMMYT, IITA, DTMA (http://dtma. cimmyt.org/index.php/varieties/dt-maize-varieties)

4.2 Contribution of Multinational Seed Companies

Multinational companies have done tremendous work for drought tolerance of maize. Significant acceleration has been made by molecular markers in attaining the higher genetic gains for yield and tolerance. MARS has doubled the rate of genetic gains in maize population of Monsanto (Eathington et al. 2007; Edgerton 2009). Regular pedigree breeding program was used by Pioneer under a program named as "mapping as you go" which also doubled the rate of genetic gains (Podlich et al. 2004). Association mapping and genomic selection have been adopted and exploited by large multinational companies more easily and quickly relative to public sector due to their extensive capability of field testing of mapping populations, synchronization of phenotyping and genotyping, massive bioinformatics resources, access to elite germplasm and capital (Eathington et al. 2007; Edgerton 2009; Schussler et al. 2011). Large multinational companies are producing about 500,000 double haploids in one calendar year which is also accelerating the speed of inbred line development for them. Seed chipping methods are also used by multinational companies for non-destructive assessment of DNA of double haploids. Unwanted DNA combinations are discarded even before sowing of seeds. However, combination of double haploids, genome-wide selection based on genotyping and seed chipping are increasing the selection pressure and shortening the number of generations needed for hybrid development which are achieving the "speeding the breeding." Multilocation testing, crop modeling, and combination of above-mentioned techniques have doubled the genetic gains for drought tolerance from 1930 to 2000 (Edgerton 2009). Drought-tolerant hybrids developed by using combination of above discussed techniques are available in market of USA since 2011. "Leading competitor hybrids" are used for yield comparison of drought-tolerant hybrids. Syngenta's Agrisure Artesian™ hybrids are better option to be used for yield comparison of drought-tolerant hybrids because these hybrids are based on series of 12 QTLs which are effective for broad range of genetic backgrounds (http://www.freepatentsonline.com/y2011/0191892.html).

Pioneer Hi-Bred (a multinational company) launched AQUAmax™ brand which comprised a line of hybrids and this was launched in 2011. For development of AQUAmax™, QTL-based approach was used which was commercially known as Accelerated Yield Technology™. Molecular mapping-based genomic selection, molecular markers as genetic covariates to highlight the genomic hotspots, and multilocation testing were the key basis of AQUAmax™ brand (Sebastian 2009). Characteristics features of AQUAmax^{TM} hybrids were having prolonged stay-green property and vigorous silking (http://www.4-traders.com/news/Pioneer-Hi-Bred-International-Inc). AQUAmax and Artesian hybrids were launched in market for sale in 2013. Drought-tolerant maize hybrids developed through conventional breeding by Pioneer Hybrid International, reported that 5 % yield was greater than drought stress, whereas drought-tolerant maize hybrids by Syngenta claimed to give 15 % more yield under drought stress (Tollefson 2011).

Monsanto is also the leading body for development of drought-tolerant transgenic maize hybrid. Droughtgard™ hybrids of Monsanto are transgenic and launched in market for sale in 2013. MON87460, a drought-tolerant maize hybrid of Monsanto has cold-shock protein gene (*cspB*); this gene was isolated from soil bacteria "*Bacillus subtilis*" (http://www.biofortified.org/2012/08/monsantos-gm-drought-tolerant-corn/). This gene encodes for a protein which act as chaperone for other proteins and this gene remain active during whole plant life and also increase the number of kernels per plant (Castiglioni et al. 2008). Recently Monsanto collaborated with BASF (Badische Anilin-und Soda-Fabrik) for drought tolerance gene discovery. Pioneer Hi-Bred and Syngenta are also working on transgenic drought-tolerant maize development.

Chapter 5
Biological Practices for Improvement of Maize Performance

Management of drought stress to reduce the yield losses in crop plants is practiced in different forms across the world. Saving irrigation water with the help of different water management practices, exploitation of agronomic practices to improve crop performance under drought stress condition, and development of drought-tolerant germplasm are the main tools which are exploited by agronomist and breeder. Water saving and cultural practices are found to be inconvenient, expensive, and requiring special skills. Development of drought-resistant germplasm is proved to be effective, efficient, and feasible approach for improving yield in drought prevailing territories (Athar and Ashraf 2009). Different strategic characters are improved by numerous biological approaches which enable the plants to escape, avoid, and tolerate the drought stress. Screening of germplasm for assessment of tolerant variants, development of tolerant genotypes through conventional breeding, mutation breeding, molecular breeding, and transgenic approaches are possible options which are working and can further be employed for further improvement. Schematic flowchart for biological or breeding strategies for improvement of drought tolerance in maize is shown in Fig. 5.1.

5.1 Screening for Drought-Tolerant Maize Germplasm

Evolutionary pathway has diverged the biological diversity at different levels of organization e.g., development of eukaryotes from prokaryotes followed by diversion toward development of plants, animals, fungi, bacteria, viruses, and other creations. Extensive prevailing diversity is categorized into different taxonomic levels like, species, genus, order, class, phylum, and kingdom. Species are further comprised of large number of varieties, strains, cultivars, and lines. Different groups of populations even within species have genetic differences for numerous parameters. Drought resistance is one of the aspects for which lot of genetic differences are present within species. These genetic differences could be assessed by screening for drought stress resistance. Earliness is critical parameter which enables the plants to escape the drought stress. Development of extensive root system and prevention of water loss enabled the plants to avoid drought stress (reviewed by Athar and

© The Author(s) 2015
M. Aslam et al., *Drought Stress in Maize (Zea mays L.)*,
SpringerBriefs in Agriculture, DOI 10.1007/978-3-319-25442-5_5

Fig. 5.1 Biological/breeding
approaches for improvement
of drought resistance in maize

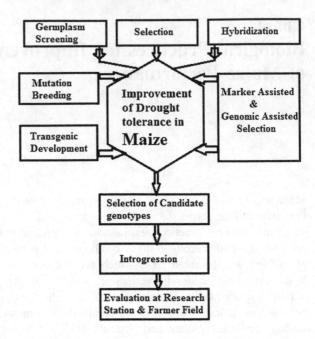

Ashraf 2009). Maintenance of normal physiological mechanisms with satisfactory yield brings drought tolerance in plants. Detail description of drought escape, avoidance, and tolerance has been discussed in Sects. 3.1, 3.2 and 3.3, respectively. Genetic variability among different genotypes of maize is present for drought escape, avoidance, and tolerance which could be retrieved by suitable screening of germplasm. Different characteristic parameters which are strongly linked with these three mechanisms are focused in screening of germplasm. Screening of maize germplasm could be done in growth room under controlled conditions and in field under natural conditions. Parameters used for screening of germplasm must be associated with grain yield because higher grain yield is ultimate objective of screening. Traits which are affected severely by drought stress are targeted in screening.

Biometrical tools assist in assessment of genetic variability among maize germplasm. Analysis of variance, mean comparison tests, basic summary statistics, metroglyph analysis, D2 statistics, principle component analysis and biplot graphical analysis are extensively used for assessment of genetic diversity (Singh and Chaudhary 1985). Drought tolerance indices are profusely used for efficient screening of germplasm in different crop plants. Stress susceptibility index (Fischer and Maurer 1978), geometric mean productivity (Fernández 1992), mean productivity (Rosielle and Hambling 1981), harmonic mean, tolerance index (Rosielle and Hambling 1981), stress tolerance index (Fernández 1992), yield index (Gavuzzi et al. 1997), yield stability index (Bouslama and Schapaugh 1984), ranking index, integrated selection index, and integrated scoring are important drought tolerance indices extensively used for effective germplasm screening. So, screening of previously available maize germplasm for drought resistance proved to be resource efficient biological strategy.

5.2 Conventional Breeding Strategies

Creation of genetic variability and novel gene combination through intercrossing of targeted parents is one of the practices used to develop tolerant genotypes. Intercrossing followed by appropriate selection scheme enables to develop an ideotype plant that is suitable for environment specific cultivation (Bänziger et al. 2000). Higher genetic variability, high heritability, and higher selection intensity empower the breeder to make appropriate selection in the germplasm (Falconer 1989). Breeding strategies for development of drought-resistant germplasm are economical and effective tool for combating the global issue of water deficiency (Subbarao et al. 2005). Existence of genetic variability at generic, specific, and varietal levels act as raw material for selection and breeding against drought stress (Serraj et al. 2005a). Maize breeders have to focus on large number of traits for improvement of drought resistance as it is well known that single trait could not improve the resistance because plant responses interact with each other in complex fashion. Gene pyramiding, efficient and systematic breeding method, can effectively improve the drought tolerance by incorporation of large number of favorable traits in one genotype. Morphological and physiological parameters which prevent water loss, improve water use efficiency, and economic yield must be focused for pyramiding as recommended by Subbarao et al. (2005). Early vigor, rapid establishment, structural and functional traits of roots, osmoprotection, stomatal conductance, and leaf characteristics are suggested by Parry et al. (2005) as key parameters for improvement of drought tolerance. Development of early maturing varieties is important tool for escaping terminal drought stress by completing the life cycle before the onset of drought stress. So, earliness could also be incorporated by breeding to escape terminal drought (Athar and Ashraf 2009). Breeding for development of genotypes which are efficient water user (collect more quantity of water and loose less) could be effective to avoid the harmful effects of drought stress. Structural and functional traits of roots and stomata should be focus of breeders for development of drought avoiding genotypes. Breeding efforts for improvement of drought tolerance concentrated on the traits which maintain normal physiological mechanisms (osmolytes, antioxidants, plant growth regulators, stress-responsive proteins, and transcription factors) and economic yield (yield and yield components) of crop plant (Bänziger et al. 2000).

Breeders collect large number of germplasm with variable origin; initially, selections (screening phase) are made on yield and yield component basis; after reducing the number of genotypes by selection then selections are made for drought resistance (testing phase). Quantitative inheritance, low heritability, and higher genotype into environment interaction proved as barriers for quick improvement of yield under drought stress (Babu et al. 2003). Assessment of yield limiting traits with the help of morphological, physiological, biochemical, and molecular techniques could supplement the conventional breeding methods for improvement of yield (Cattivelli et al. 2008).

Conventional breeding methods rely on conducting multilocation, multiyear and multiseason yield trials for evaluation of stability in the performance against drought stress (Babu et al. 2003). Yield and yield components are primary target traits to be focused for crop improvement against drought stress. Secondary traits are equally important for breeding against drought stress. Secondary traits which have strong correlation with grain yield, stable in nature, easy to measure, high heritability, and improve yield are preferred even under normal environmental conditions (Edmeades et al. 2001). Worth of secondary traits is realized through selection indices, heritability in progenies, and genetic association. Development of near isogenic lines and synthetics helped the breeder to know the association of targeted secondary trait with economic yield (Bänziger et al. 2000). From practical perception of breeder, secondary traits which are important for improvement of drought tolerance and recommended by CIMMYT for use in breeding programs have been enlisted in Table 5.1.

Interspecific and intervarietal differences are present for water use efficiency in different crop plants and impairment of this trait is among early drought responses. Water use efficiency is genetically governed trait and its higher value depicts

Table 5.1 Secondary traits targeted for drought tolerance improvement through conventional breeding (Recommended by CIMMYT; Bänziger et al. 2000)

Trait	Heritability	Correlation with yield	Selection objective	Target growth stage
Grain yield	Medium to low under flowering stress, medium during grain filling stress	High positive	Increase economic yield	Flowering and grain development
Ears per plant	High and increasing with stress intensity	High	More ears per plant or low barrenness	Flowering stage
Anthesis-silking interval (ASI)	Medium under normal, high level under severe stress	High under stress	Reduced or negative ASI	Flowering stage
Tassel size	Medium to high	Medium	Smaller tassel with fewer branches	Could be measured under normal and stress conditions
Leaf senescence	Medium	Medium under grain filling stress	Delayed leaf senescence or stay-green property	Grain filling stage
Leaf rolling	Medium to high	Medium to low	Unrolled leaves	Flowering stress

drought tolerance. Water use efficiency is reduced under drought stress either due to sustained biomass production or higher water losses. Higher ratio for biomass production to transpired water, partitioning of biomass toward economical part, reduced water loss, and increased water uptake are the components of water use efficiency which could be focused for breeding against drought stress (Condon et al. 2004; Farooq et al. 2009).

Improvement in drought tolerance is complex due to polygenic nature and low frequency of alleles for tolerance in maize. Open-pollinated varieties (OPV) and hybrid products are targets in maize for improvement of drought tolerance. Improving locally adapted germplasm, improving tolerant exotic germplasm for adaptability, and development of new breeding population through introgression are the recommended options for drought tolerance improvement in maize. Development of source population and evaluation of that population are subcomponents of introgression. Selection and development of source population must be done on the basis of following characteristics; general adaptability, grain color, grain texture, maturity, disease resistance, abiotic stress tolerance, heterotic pattern, heterotic response, combining ability, and other value added traits. Evaluation of developed population could be done through line evaluations, hybridization, diallels of local or exotic populations or lines, population x local tester topcross, and line x local tester topcross. Intrapopulation improvement for drought tolerance could be done through individual plant selection, per se performance, test crosses using individual plants, half-sib progenies and parental testers (Bänziger et al. 2000).

Mutation breeding is very important component of crop breeding program that is involved in exploitation of mutations for improvement in agricultural and horticultural crops. Chemical and physical mutagenic agents are used for induction of mutations. Mutants are used either direct as new cultivars or as parent for development of new cultivars (Waugh et al. 2006). Mainly improvement is brought by mutation breeding through up-gradation of well-adapted genotypes which are deficit for one or few traits (Wilde et al. 2012). Increase in genetic variability by developing novel alleles, variety development in 2–3 generations, and generation of chimera in somatic tissues are the characteristic benefits of mutation breeding (Roychowdhury and Tah 2013). Seed, pollen, whole plant, tubers, cuttings, bulbs, corms, stolons, tissues, and suspended cells could be used as plant material for mutagenesis.

FAO/IAEA has developed a database called "Mutant Varieties Database" (http://www-mvd.iaea.org). Details of all varieties developed through mutation breeding, breeding methods, primary and secondary traits improved, year of registration, and country of origin have been provided in this database. International Atomic Energy Association (IAEA) has categorized the mutant database based on breeding methods; (a) mutants directly used as commercial cultivars, (b) mutants used as one parent in hybridization program, (c) commercial cultivars developed using both mutant parents, (d) hybrid development using one mutant parent, and (e) commercial cultivars developed through mutation of segregating populations. Aberrant lateral root formation, unusual gravitropism behavior, lack of crown and brace roots, and premature root degradation are observed by mutation of maize roots (Feix et al. 1997).

Improvement in technology furnished mutation breeding with many high-throughput techniques like, TILLING (Targeting Induced Limited Lesions IN Genomes), EcoTILLING and high-resolution melt analysis (HRM). Efficiency and efficacy of mutation breeding in crop breeding increased with the help of molecular mutation techniques. TILLING is efficient reverse genetics tool which comprised mutagenic treatment followed by detection of point mutations with the help of sophisticated detection tools. Mutagenic treatment, development of segregating population (M2), sample collection and their DNA extraction, pooling samples (8–12), identification of induced point mutation, and validation followed by evaluation of identified mutants are the key steps in the process of TILLING. Mutants developed and identified through TILLING could be used in breeding programs and gene-function assessment. TILLING was used for induction and identification of point mutations in numerous crops. Maize pollen population was mutagenized followed by detection of mutations through TILLING tools. DNA segment of 1 kb was pooled from 11 different genes and 17 different-independent-induced point mutations were obtained (Roychowdhury and Tah 2013).

Alternative TILLING approach, known as EcoTILLING, is effective tool for identification of SNPs in natural populations which has been induced by sponta-neous mutations. Point mutations, deletions, and insertion within target sequence can be determined by EcoTILLING (Roychowdhury and Tah 2013). High-resolution melt analysis (HRM) is an alternative screening approach which is used for detection of mutations in genes that have multiple exons and introns. Single-base mismatching is detected and effectively used for genotyping in medical, single-nucleotide polymorphism (SNP) discovery and SNP genotyping in plants (Zhou et al. 2004, 2005).

It is well admitted that conventional breeding approaches are effectively involved in the improvement of drought tolerance in maize. These techniques are effective because evaluation is made under field conditions and interaction with environment is well considered. Mutation-assisted breeding alongwith modern molecular techniques are effective in generation and identification of desired mutations. Mutation breeding being nontransgenic approach is safe and secure strategy for crop improvement. Ideotype for drought-tolerant maize genotype can be developed through mutation-assisted breeding, so there is need to explore the potential of mutations for maize improvement.

5.3 Marker-Assisted and Genomic-Assisted Breeding

Effectiveness of conventional breeding is reduced due to low heritability of traits in field, high-field management cost, seasonal variability, time and space issues and higher genotype into environment interaction. So marker-assisted selection becomes effective tool because DNA present within cell is independent of environmental and managerial effects. Pace of crop improvement increases by marker-assisted breeding because these are based on cellular DNA. Numerous DNA markers are available but plant breeder need ideal markers which must have following characteristics features and must be strongly linked with trait controlling genes, co-dominant inheritance, and PCR based; marker should depict large variability for traits, polymorphic in nature, abundant in genome, and easy to amplify (Varshney 2010). Marker-assisted selection is used for numerous tasks. Genetic distance between parents, prediction of heterotic potential of hybrids, and selection of inbred lines to be used as parent could be made effectively by fingerprinting of inbred lines. Line conversion in maize can also be done through marker-assisted backcrossing by transferring the one desired trait coding gene from donor line to recipient line. Number of generations for backcrossing and probability of linkage drag is reduced significantly in cases of marker-assisted backcrossing comparative to conventional backcrossing (Bänziger et al. 2000).

Tolerance is complex feature governed by large number of traits, and these traits are controlled by large number of chromosomal regions known as quantitative trait loci (QTLs). Parents with contrasting phenotypic expression are crossed to develop segregating progenies. Segregating populations are screened with the help of DNA markers like, RAPD, RFLP, AFLP, SSR, and SNPs. Markers linked with specific traits are then identified with bioinformatics tools. Exploitation of DNA-based markers for identification of QTL mapping linked with morphological, physiological, and biochemical traits could be targeted by breeder for drought resistance improvement. After identification of QTLs linked with traits, drought tolerance can be improved by introgression of these QTLs into modern promising cultivars. Marker-assisted selection (MAS) based on trait-linked QTLs proved to be effective for dissecting quantitative traits into unit genetic components and assisting plant breeder to make appropriate-targeted selection (Chinnusamy et al. 2005; Hussain 2006).

Linkage mapping and association studies through association mapping and candidate gene approach are effective for identification of QTLs. Most of QTL studies in literature are based on segregation mapping but association mapping is most vigorous tool than segregation mapping (Syvänen 2005). Monogenic traits like plant height, osmotic adjustment, flowering time, and ear development are more adaptive traits for drought tolerance. Genetic diversity for numerous morphological, physiological, biochemical, and molecular drought responsive traits in maize is reported. So, genetic variability could be exploited for improvement of drought tolerance in maize through marker-assisted selection. MAS proved even more efficient tool if markers are strongly linked with stress-responsive traits. Anthesis silking interval (ASI) is very important trait in maize; lower ASI value is associated with drought tolerance. CIMMYT identified six QTLs linked with ASI, which are located on chromosome number 1, 2, 5, 6, 8, and 10 of maize genome. These QTLs are contributing 50 % phenotypic variability of ASI and are stable across the years and water regimes (Bänziger et al. 2000). Additive QTLs for ear length and kernel weight, whereas epistatic QTLs for kernel number per row are observed in maize recombinant inbred lines (RILs). Genetic background of QTLs is changed (additive to apistatic and vice versa) under different water treatments for some maize traits.

Existence of additive and epistatic QTLs in maize shows that expression pattern of traits is diverse and nature of drought tolerance is very complex (Lu et al. 2006).

Information obtained from MAS made the breeding programs more effective by following ways; selection can be made at early generations and number of generations required for conventional breeding approach are reduced; accuracy of selection is highly increased (Phelps et al. 1996). Depending on the targeted locus, site of amplification, level of conservation, type of primers used, and breeding objectives large number of DNA-based markers are being used, some of them are enlisted here: random amplified polymorphic DNA (RAPD), selective amplification of microsatellite polymorphic loci (SAMPL), restriction fragment length polymorphism (RFLP), sequence characterized amplified regions (SCAR), expressed sequence tags (EST), simple sequence repeats (SSR), inter-simple sequence repeat (ISSR), single nucleotide polymorphism (SNP), sequence specific amplification polymorphisms (S-SAP), sequence tagged site (STS), sequence tagged microsatellite site (STMS), single-primer amplification reactions (SPAR), site-selected insertion PCR (SSI), single-stranded conformational polymorphism (SSCP), short-tandem repeats (STR), diversity arrays technology (DART), and variable number tandem repeat (VNTR) (Semagn et al. 2006).

Research efforts in maize were focused on the development of microsatellite markers for germplasm analysis and genetic mapping. Gene mapping is helpful in providing the information about specific locus of genes and number of genes governing the traits. Dubey et al. (2009) targeted the 24 accessions of tropical maize for assessment of drought-linked SSR markers. They found that UMC1042, DUPSSR12, UMC1056, BNLG1866, UMC1069, DUP13, BNLG1028, UMC1962, and C1344 SSR markers were linked with drought responses. Tuberosa et al. (2002a) and Sawkins et al. (2006) identified the QTLs in maize, which were linked with drought. Introgression breeding in maize, introgression of transgenes, conversion of simple or complex traits, and marker-assisted recurrent selection (MARS) were breeding perspectives in maize for which markers were used with special focus to drought stress (Ragot et al. 1995; Hospital et al. 1997; Sawkins et al. 2006).

Theoretically MAS is known to improve drought tolerance but practically contribution of MAS in release of high yielding drought-tolerant genotype is nonsignificant (Reynolds and Tuberosa 2008). So, focus should be targeted that markers linked with drought tolerance should also be linked with higher yield potential for getting two fold benefits. In literature, few cases had been reported which showed the involvement of MAS in development of drought tolerance cultivars with higher yield potential (Reviewed by Athar and Ashraf 2009). Ribaut and Ragot (2007) mentioned that introgression of five QTLs in maize has increased 50 % yield comparative to standard hybrids under drought stress, and no yield losses were observed under normal water availability. Introgression of yield-linked QTLs in pearl millet improved the grain yield in drought sensitive genotypes (Serraj et al. 2005b). Stay-green character of sorghum is improved by introgression of QTLs (Harris et al. 2007).

QTL mapping enabled the breeders to identify the chromosomal regions linked with different plant traits. Effect of genetic background, complex genetic basis, stage of plant growth and development, environment × QTL interaction (Tuberosa et al. 2002b), gene by gene effects, inadequate phenotyping, cost, and skill issues are limiting the effectiveness of QTL mapping (Campos et al. 2004; Xu, et al. 2009). QTL identification, validation in different populations, or under different environments followed by their proper manipulation in breeding program could be much more effective for real-sense improvement against drought stress. There is still gap which must be filled for getting more benefits from marker-based selection.

Functional genomics and transcriptomics are recently used for extensive understanding of plant responses against stresses. Identification of candidate gene, followed by characterization, and determination of transcriptomic responses through microarray or whole genome sequencing help to clearly highlight the tolerance mechanisms. Drought responsive candidate genes are identified by imposition of drought stress on stress-responsive genotypes followed by ESTs generation from either normalized or nonnormalized cDNA library. Public data bases are being exploited for retrieval of drought stress-responsive candidate genes in major crops like wheat, maize, barley, and rice (Sreenivasulu et al. 2007; Kathiresana et al. 2006).

Transcript profiling is used for identification of candidate genes through assessment of differential gene expressions in a specific tissue at different times (Hampton et al. 2010). Transcript profiling can be done through cDNA–amplified fragment length polymorphism (cDNA–AFLP), PCR-based differential display PCR (DDRT-PCR) analysis, digital expression analysis based on counts of ESTs, cDNA and oligonucleotide microarrays, serial analysis of gene expression (SAGE) technique, SuperSAGE, and next generation sequencing (reviewed by Mir et al. 2012). Among these techniques, next generation sequencing (NGS)-based techniques are most preferred for routine transcript profiling of main crops for identification of drought-tolerant candidate gene followed by exploitation of that gene through genomics and marker-assisted breeding. After identification of major QTLs, contributing to drought tolerance, these are validated in target population. Following validation, these QTLs could be exploited through their introgression into high yielding and drought susceptible (recipient parent) from low yielding and drought-tolerant parent (donor parent), this technique is known as marker-assisted backcrossing (MABC). Birsa Vikas Dhan 111 (PY 84), a rice variety, was developed through marker-assisted backcrossing in India having improved drought tolerance (Steele et al. 2007). Complexity of mechanism of drought tolerance act as barrier for substantial exploitation of MABC e.g., almost 10 % phenotypic variability was explained by identified QTLs in maize (Xu et al. 2009). These findings conclude that extensively large population size is mandatory for achieving satisfactory improvement through MABC.

MABC acts as effective tool when traits are governed by single or few genes but in case of drought tolerance which is very complex feature and governed by large number of genes, this technique becomes least effective. Marker-assisted recurrent selection (MARS) has capability to deal with complex traits like drought tolerance and involves the inter-mating of selected accessions in each recurrent cycle (Ribaut and Ragot 2007). Population improvement is adequately accomplished by MARS because MAS is practiced in each selection cycle followed by interbreeding of selected individuals, which validates and increases the frequency of desired genes in target population (Eathington et al. 2007). MARS is being used for drought tolerance improvement in different crops e.g., wheat, chickpea, cowpea, and sorghum (reviewed by Mir et al. 2012). Plenty of work needed to be done for improvement of drought tolerance in maize through MARS.

Genome selection (GS) and genome-wide selection (GWS) are important molecular techniques for improvement of drought tolerance in crop plants by developing superiorly drought-tolerant genotypes. Unlikely of MARS, genome selection is done through genome-wide marker genotyping. Breeding methodology for GS was described by Meuwissen et al. (2001) and Mir et al. (2012). GS has numerous advantages over other techniques like, reduced selection time, increased annual gain from selection, and reduced phenotyping frequency (Rutkoski et al. 2010). Initiative for exploration of GS potential in different crops has been taken but its application for improvement of drought tolerance (Mir et al. 2012) especially in maize is lacking. Exploration of GS in maize for the improvement of drought tolerance is recommended.

5.4   Transgenic Maize Development

Genetically complex nature of drought tolerance makes the transgenic development even more complex than for monogenic traits. Exploitation of signal transduction cascades, transcription factors, or transformation with numerous genes regulates the pivotal processes. But current research work is focused on single-gene transformation. Signal transduction pathways are activated by stress responses which resultantly regulate the cascades of adaptations. These signaling pathways can be modified or tailored using tool of genetic engineering. Development of drought-tolerant transgenic crop basically involves the incorporation of one or more genes from other donor source/sources in target crop to modify the signaling and subsequent events (Vinocur and Altman 2005). Genes which can be manipulated in genetic engineering are categorized into four main classes: (1) genes involved in transcriptional and signal transduction pathways, (2) genes involved in protection of cellular membranes and biosynthesis of stress-responsive proteins, (3) genes involved in uptake of ions and water-like ion transporters and aquaporins (Wang et al. 2003), and (4) genes involved in cellular metabolism e.g., free amino acids, proline, soluble sugars, polyols, and glycinebetaine (Vinocur and Altman 2005).

Transgenic constitutive upregulation of transcription factors (TFs) improve drought tolerance but these TFs also upregulate other genes which impair normal plant growth and development resultantly reduced economic yield (Wang et al. 2003). Alternative to TFs, stress-induced promoters could be exploited for improvement of drought tolerance because their side effects are far less than TFs (Athar and Ashraf 2009). NADP-malic enzyme, key enzyme of C4 photosynthesis, from maize was transgened in tobacco which reduced stomatal conductance and improved water use efficiency (Laporte et al. 2002). So functional sustainability of this enzyme in maize under drought stress definitely will improve the tolerance by improving water use efficiency. Transgenic plants for improved drought were developed in arabidopsis, rice, wheat, tobacco, tomato, brassicas, and others (Athar and Ashraf 2009). Monsanto developed drought-tolerant transgenic maize genotype known as MON87460 which has permission for sale in United States Department of Agriculture (Gilbert 2010). Monsanto and BASF developed another maize transgene known as DroughtGuard Hybrid Corn which carried the cold shock protein (CSPB) from *Bacillus subtilis*. Results of multilocation yield trails showed that yield of transgenic maize was 5774.5 kg/ha and yield of nontransgenic standard was 4770.25 kg/ha. mtID (bacterial mannitol-1-phosphate dehydrogenase) and HVA1 (*Hordeum vulgare)* pyramiding in maize improved the drought tolerance (Nguyen et al. 2013). Mitogen-activated protein kinase kinase kinase (MAPKKK) is involved in conferring tolerance against different abiotic stresses. MAPKKK from tobacco called NPK1 was transgened in maize which improved drought tolerance by improving photosynthesis rate and grain weight under drought stress relative to nontransgenic maize (Shou et al. 2004).

Theoretically, it is possible to improve drought tolerance by developing transgenic crop plants but practically there are many limitations which reduce the effectiveness of transgenic crops. Side effects of transgenes and complexity of tolerance mechanism make improvement very difficult (Cattivelli et al. 2008). Suitability of transgene, dosage effect, level of tolerance, side effects of transgene, yield penalty and socio-scientific acceptance of transgenics in food crops like, maize, are determinants for effectiveness of transgenic development.

Chapter 6
Conclusions and Summary

Drought is major abiotic stress which hinders crop productivity across the world. Drought affects numerous crop plants at different levels of growth and development. Maize is important cereal crop and grown in large number of countries across the world. Effects of drought stress on maize are prevalent from germination to harvest maturity. Germination percent, germination potential, germination rate, seedling establishment and seedling vigor are disturbed by drought at early growth stages. Growth and development of vegetative parts of maize are seriously affected by diminished cell division and cell proliferation which clarified that cell cycle is critically dependent on water status of plants. Plant height, stem diameter, plant biomass, leaf area and root development are disturbed in maize by drought stress. Reproductive stage in maize is more critically impaired by drought stress. Development of tassel and ear, pollination, fertilization, embryo development, endosperm development and grain filling are seriously affected by drought stress in maize.

All genotypes of maize are not equally affected by drought stress due to high level of variability in genetic background of this crop. Different mechanisms have been evolved in maize like other crops which enable them to effectively survive under drought stress. Drought escape, drought avoidance and drought tolerance are different mechanisms which work under the heading of drought resistance. These mechanisms are evolved in different maize genotypes through course of evolution and domestication. Drought escaper maize genotypes modulate their life to be completed before the onset of drought and drought avoider maize genotypes avoid themselves from drought either by reducing water losses or by increasing water uptake. Drought tolerant maize genotypes maintain their growth and development along with economical grain yield under drought stress. On the other hand, drought susceptible maize genotypes may be lacking any one of these adaptive mechanisms and facing severe damage in terms of growth, development and grain yield. Drought tolerance is very complex mechanism which is collaboratively conferred by osmotic adjustment, plant growth regulators, antioxidative defense, stress responsive proteins, water channel proteins, transcription factors and signal transduction pathways.

CIMMYT, IITA, Monsanto, Syngenta and Pioneer are the leading maize research groups in the world. These groups have worked and working on numerous

© The Author(s) 2015
M. Aslam et al., *Drought Stress in Maize (Zea mays L.)*,
SpringerBriefs in Agriculture, DOI 10.1007/978-3-319-25442-5_6

aspects of maize crop however, drought tolerance is also one of their key research objectives. Numerous research projects have been completed in which different breeding methods were practiced for improvement of drought tolerance in maize. Drought tolerant maize hybrids and OPVs are practically cultivated in numerous African countries whoever, Monsanto has developed drought tolerant transgenic maize.

As demand of maize is increasing day by day and water scarcity is increasing so, there is dire need to further improve the level of drought resistance in maize. Different strategies are used for improvement of maize against drought stress e.g. managerial strategies and biological strategies. Managerial strategies involve the usage of water resources and adoption of water saving agronomic practices. On the other hand biological approaches deal with manipulation of genetic background of maize for improvement against drought stress. Biological strategies are preferred over managerial practices due to long term and economical effectiveness. Biological approach is practiced in different forms. Available germplasm has lot of genetic variability which can be exploited for higher drought tolerance through effective screening tools. Conventional breeding, mutation breeding, marker assisted and genomic assisted breeding and development of drought tolerant transgenic maize are numerous strategies which are enlisted under the heading of biological strategies. There are few gaps in effectiveness of biological techniques which must be filled under the umbrella of modern technology for the improvement of drought tolerance of maize in such a way that it can fully combat with drought stress. Lots of novel breeding and evaluation techniques have been developed in recent past and practical application of other techniques will further help to cope the problem of drought stress by development of more drought tolerant maize genotypes for resolution of food security.

References

Abdelgawad ZA, Khalafaallah AA, Abdallah MM (2014) Impact of methyl jasmonate on antioxidant activity and some biochemical aspects of maize plant grown under water stress condition. Agric Sci 5:1077–1088

Abebe T, Guenzi AC, Martin B, Cushman JC (2003) Tolerance of mannitol-accumulating transgenic wheat to water stress and salinity. Plant Physiol 131:1748–1755

Achakzai AKK (2009) Effect of water stress on imbibition, germination and seedling growth of maize. Sarhad J Agric 25(2):165–172

Aina PO, Fapohunda HO (1986) Root distribution and water uptake patterns of maize cultivars field-grown under differential irrigation. Plant Soil 94:257–265

Ali Q, Ashraf M (2011) Exogenously applied glycinebetaine enhances seed and seed oil quality of maize (*Zea mays* L.) under water deficit conditions. Environ Exp Bot 71:249–259

Allakhverdiev SI, Hayashi H, Nishiyama Y, Ivanov AG, Aliev JA, Klimov VV, Murata N, Carpemtier R (2003) Glycinebetaine protects the D1/D2/Cyt *b* 559 complex of photosystem II against photo-induced and heat-induced inactivation. J Plant Physiol 160:41–49

Almansouri M, Kinet JM, Lutts S (2001) Effect of salt and osmotic stresses on germination in durum wheat (*Triticum durum* Desf.). Plant Soil 231:243–254

Amara I, Odena A, Oliveira E, Moreno A, Masmoudi K, Page's M, Goday A (2012) Insights into maize LEA proteins: from proteomics to functional approaches. Plant Cell Physiol 53(2):312–329

Andersen MN, Asch F, Wu Y, Jensen CR, Naested H, Mogensen VO, Koch KE (2002) Soluble invertase expression is an early target of drought stress during the critical, abortion sensitive phase of young ovary development in maize. Plant Physiol 130:591–604

Anjum SA, Wang LC, Farooq M, Hussain M, Xue LL, Zou CM (2011a) Brassinolide application improves the drought tolerance in maize through modulation of enzymatic antioxidants and leaf gas exchange. J Agron Crop Sci 197:177–185

Anjum SA, Xie X-Y, Wang L-C, Saleem MF, Man C, Lei W (2011b) Morphological, physiological and biochemical responses of plants to drought stress. Afr J Agric Res 6 (9):2026–2032

Anjum SA, Saleem MF, Wang L-C, Bilal MF, Saeed A (2012) Protective role of glycinebetaine in maize against drought induced lipid peroxidation by enhancing capacity of antioxidative system. Aust J Crop Sci 6(4):576–583

Apel K, Hirt H (2004) Reactive oxygen species: metabolism, oxidative stress, and signal transduction. Annu Rev Plant Biol 55:373–399

Araus JL, Slafer GA, Reynolds MP, Royo C (2002) Plant breeding and drought in C3 cereals: what should we breed for? Ann Bot 89:925–940

Artlip TS, Madison JT, Setter TL (1995) Water deficit in developing endosperm of maize: cell division and nuclear DNA endoreduplication. Plant Cell Environ 18:1034–1040

Arve LE, Torre S, Olsen JE, Tanino KK (2011) Ch#12: Stomatal responses to drought stress and air humidity. In: Shanker A, Venkateswarlu B (eds) Abiotic stress in plants mechanisms and adaptations. InTech Publishers. doi:10.5772/24661

© The Author(s) 2015

M. Aslam et al., *Drought Stress in Maize (Zea mays L.)*,
SpringerBriefs in Agriculture, DOI 10.1007/978-3-319-25442-5

Ashraf M, Foolad MR (2007) Roles of glycinebetaine and proline in improving plant abiotic stress resistance. Environ Exp Bot 59:206–216

Aslam M, Maqbool MA, Zaman QU, Latif MZ, Ahmad RM (2013a) Responses of Mungbean genotypes to drought stress at early growth stages. Int J Basic Appl Sci IJBAS IJENS 13 (05):23–28

Aslam M, Basra SMA, Maqbool MA, Bilal H, Zaman QU, Bano S (2013b) Physio-chemical distinctiveness and metroglyph analysis of cotton genotypes at early growth stage under saline hydroponics. Int J Agric Biol 15:1133–1139

Aslam M, Maqbool MA, Akhtar S, Faisal W (2013c) Estimation of genetic variability and association among different physiological traits related to biotic stress (*Fusarium Oxysporum* L.) in chickpea. J Anim Plant Sci 23(6):1679–1685

Aslam M, Ahmad K, Maqbool MA, Bano S, Zaman QU, Talha GM (2014a) Assessment of adaptability in genetically diverse chickpea genotypes (*Cicer arietinum* L.) based on different physio-morphological standards under ascochyta blight inoculation. Int J Adv Res 2(2):245–255

Aslam M, Zeeshan M, Maqbool MA, Farid B (2014b) Assessment of drought tolerance in maize (*Zea May* L.) genotypes at early growth stages by using principle component and biplot analysis. Exp 29(1):1943–1951

Aslam M, Zamir MSI, Afzal I, Amin M (2014c) Role of potassium in physiological functions of spring maize (*Zea Mays* L.) grown under drought stress. J Anim Plant Sci 24(5):1452–1465

Association of Official Seed Analysts (AOSA) (2002) Seed vigor testing handbook. (Contribution, 32): Stillwater.

Athar HR, Ashraf M (2009) Strategies for crop improvement against salinity and drought stress: an overview. In: Ashraf M, Ozturk M, Athar HR (eds) Salinity and water stress. Springer Science, Heidelberg

Aylor DE (2002) Settling speed of corn (Zea mays) pollen. J Aerosol Sci 33:1599–1605

Aylor DE (2004) Survival of maize (Zea mays) pollen exposed in the atmosphere. Agric For Meteorol 123:125–133

Babu RC, Nguyen BD, Chamarerk VP, Shanmugasundaram P, Chezhian P, Jeyaprakash SK, Ganesh A, Palchamy S, Sadasivam S, Sarkarung S, Wade LJ, Nguyen HT (2003) Genetic analysis of drought resistance in rice by molecular markers. Crop Sci 43:1457–1469

Bahrun A, Jensen CR, Asch F, Mogensen VO (2002) Drought-induced changes in xylem pH, ionic composition, and ABA concentration act as early signals in field grown maize (*Zea mays* L.). J Exp Bot 53(367):251–263

Bajguz A (2007) Metabolism of brassinosteroids in plants. Plant Physiol Biochem 45:95–107

Bajji M, Lutts S, Kinet JM (2001) Water deficit effects on solute contribution to osmotic adjustment as a function of leaf ageing in three durum wheat (*Triticum durum* Desf.) cultivars performing differently in arid conditions. Plant Sci 160:669–681

Bano A, Yasmeen S (2010) Role of phytohormones under induced drought stress in wheat. Pak J Bot 42:2579–2587

Bänziger M, Edmeades GO, Beck D, Bellon M (2000) Breeding for drought and nitrogen stress tolerance in maize: from theory to practice. CIMMYT, Mexico

Bänziger M, Setimela PS, Hodson D, Vivek B (2006) Breeding for improved drought tolerance in maize adapted to Southern Africa. Agric Water Manag 80:212–224

Bänziger M, Araus J (2007) Recent advances in breeding maize for drought and salinity stress tolerance. In: Jenks MA, Hasegawa PM, Jain SM (eds) Advances in molecular breeding toward drought and salt tolerant crops. Springer, Netherlands, pp 587–601

Barker TC, Campos H, Cooper M, Dolan D, Edmeades GO, Habben J, Schussler J, Wright D, Zinselmeier C (2005) Improving drought tolerance in maize. Plant Breed Rev 25:173–253

Barow M, Meister A (2003) Endopolyploidy in seed plants is differently correlated to systematics, organ, life strategy and genome size. Plant Cell Environ 26:571–584

Bassetti P, Westgate ME (1993) Water deficit affects receptivity of maize silks. Crop Sci 33:279–282

Battaglia M, Olvera-Carrillo Y, Garciarrubio A, Campos F, Covarrubias AA (2008) The enigmatic LEA proteins and other hydrophilins. Plant Physiol 148:6–24

Becker TW, Fock HP (1986) Effects of water stress on the gas exchange, the activities of some enzymes of carbon and nitrogen metabolism, and on the pool sizes of some organic acids in maize leaves. Photosynth Res 8:175–181

Belaygue C, Wery J, Cowan AA, Tardieu F (1996) Contribution of leaf expansion, rate of leaf appearance, and stolon branching to growth of plant leaf area under water deficit in white clover. Crop Sci 36:1240–1246

Bernardo R (2008) Molecular markers and selection for complex traits in plants: learning from the last 20 years. Crop Sci 48:1649–1664

Bianchi A, Bianchi G, Avato P, Salamini F (1985) Biosynthetic pathways of epicuticular wax of maize as assessed by mutation, light, plant age and inhibitor studies. Maydica 30:179–198

Blum A (1988) Plant breeding for stress environment. CRC Press Inc, Boca Raton

Blum A (2011) Plant breeding for water-limited environments. Springer, London

Bohnert HJ, Nelson DE, Jensen RG (1995) Adaptations to environmental stresses. *Plant* soluble sugar contents of *Sorghum bicolor* (L.) Moench seeds under various abiotic stresses. Plant Growth Reg 40:157–162

Bolaños J, Edmeades GO (1996) The importance of the anthesis-silking interval in breeding for drought tolerance in tropical maize. Field Crops Res 48:65–80

Bouslama M, Schapaugh WT (1984) Stress tolerance in soybean. Part 1: evaluation of three screening techniques for heat and drought tolerance. Crop Sci 24:933–937

Boyle MG, Boyer JS, Morgan PW (1991) Stem infusion of liquid culture medium prevents reproductive failure of maize at low water potential. Crop Sci 31:1246–1252

Gambín BL, Borrás L, Otegui ME (2007) Kernel water relations and duration of grain filling in maize temperate hybrids. Field Crops Res 101:1–9

Brown RF, Mayer DG (1988) Representing cumulative germination. 1. A critical analysis of single-value germination indices. Ann Bot 61:117–125

Bruce WB, Edmeades GO, Barker TC (2002) Molecular and physiological approaches to maize improvement for drought tolerance. J Exp Bot 53:13–25

Buitink J, Walters-Vertucci C, Hoekstra FA, Leprince O (1996) Calorimetric properties of dehydrating pollen; analysis of desiccation tolerant and an intolerant species. Plant Physiol 111:235–242

Burris JS (2001) Adventitious pollen intrusion into hybrid maize seed production fields. In: Proceedings of the 56th annual Corn and Sorghum research conference. American Seed Trade Association Inc, Washington

Butzen S, Schussler J (2009) Pioneer research to develop drought-tolerant corn hybrids. Crop Insights 19, No 10. Pioneer HiBred Intl, Des Moines, 4 p

Cahn MD, Zobel RW, Bouldin DR (1989) Relationship between root elongation rate and diameter and duration of growth of lateral roots of maize. Plant Soil 119:271–279

Cairns JE, Crossa J, Zaidi PH, Grudloyma P, Sanchez C, Araus JL, Thaitad S, Makumbi DC, Magorokosho M, Bänziger Menkir A, Hearne S, Atlin GN (2013) Identification of drought, heat, and combined drought and heat tolerant donors in maize (*Zea mays* L.). Crop Sci 3:1335–1346

Campos H, Cooper A, Habben JE, Edmeades GO, Schussler JR (2004) Identification of quantitative trait loci under drought conditions in tropical maize. 1. Flowering parameters and the anthesis-silking interval. Theor Appl Genet 92:905–914

Castiglioni P, Warner D, Bensen RJ, Anstrom DC, Harrison J, Stoecker M, Abad M, Kumar G, Salvador S, D'Ordine R, Navarro S, Back S, Fernandes M, Targolli J, Dasgupta S, Bonin C, Luethy M, Heard JE (2008) Bacterial RNA chaperones confer abiotic stress tolerance in plants and improved grain yield in maize under water-limited conditions. Plant Physiol 147:446–455

Castleberry RM, Crum CW, Krull CF (1984) Genetic improvements of U.S. maize cultivars under varying fertility and climatic environments. Crop Sci 24:33–36

Cattivelli L, Rizza F, Badeck FW, Mazzucotelli E, Mastrangelo AM, Francia E, Mare C, Tondelli A, Stanca AM (2008) Drought tolerance improvement in crop plants: an integrative view from breeding to genomics. Field Crop Res 105:1–14

Ceron-Garcia A, Vargas-Arispuro I, Aispuro-Hernandez E, Martinez-Tellez MA (2012) Ch#1: Oligoglucan elicitor effects during plant oxidative stress. In: Bubulya P (ed) Cell metabolism-cell homeostasis and stress response. InTech Publishers. doi:10.5772/26057

CGIAR Generation Challenge Programme (2014) 2014 Project Updates (incorporating projects completed in 2013). Generation Challenge Programme, Texcoco

Chaves MM, Flexas J, Pinheiro C (2009) Photosynthesis under drought and salt stress: regulation mechanisms from whole plant to cell. Ann Bot 103:551–560

Chaves MM, Pereira JS, Maroco J, Rodrigues ML, Ricardo CPP, Osório ML, Carvalho I, Faria T, Pinheiro C (2002) How plants cope with water stress in the field. Photosynth Growth Ann Bot 89:907–916

Chen M, Wang QY, Cheng XG, Xu ZS, Li LC, Ye XG, Xia LQ, Ma YZ (2007) GmDREB2, a soybean DRE-binding transcription factor, conferred drought and high-salt tolerance in transgenic plants. Biochem Biophys Res Commun 353:299–305

Chen THH, Murata N (2008) Glycinebetaine: an effective protectant against abiotic stress in plants. Trends Plant Sci 13:499–505

Cheng W-H, Taliercio EW, Chourey PS (1996) The Miniature1 seed locus of maize encodes a cell wall invertase required for normal development of endosperm and maternal cells in the pedicel. Plant Cell 8:971–983

Chinnusamy V, Jagendorf A, Zhu JK (2005) Understanding and improving salt tolerance in plants. Crop Sci 45:437–448

Chugh V, Kaur N, Gupta AK (2011) Evaluation of oxidative stress tolerance in maize (*Zea mays* L.) seedlings in response to drought. Indian J Biochem Biophys 48(1):47–53

Chugh V, Kaur N, Grewal MS, Gupta AK (2013) Differential antioxidative response of tolerant and sensitive maize (*Zea mays* L.) genotypes to drought stress at reproductive stage. Indian J Biochem Biophys 50(2):150–158

Claassen MM, Shaw RH (1970) Water deficit in corn 1. Vegetative component. Agron J 62:649–652

Condon AG, Richards RA, Rebetzke GJ, Farquhar GD (2004) Breeding for high water-use efficiency. J Exp Bot 55:2447–2460

Crafts-Brandner SJ, Salvucci ME (2002) Sensitivity of photosynthesis in a C4 plant, maize, to heat stress. Plant Physiol 129:1773–1780

Cristiana F, Nina Z, Elena A (2012) Homocysteine in red blood cells metabolism pharmacological approaches. In: Moschandreou TE (ed) Blood cell—an overview of studies in hematology. InTech Publisher. doi:10.5772/47795.

Dass S, Arora P, Kumari M, Pal D (2001) Morphological traits determining drought tolerance in maize (*Zea mays* L.). Indian. J Agric Res 35(3):190–193

Davies WJ, Tardieu F, Trejo CL (1994) How do chemical signals work in plants that grow in drying soil? Plant Physiol 104:309–314

Delachiave MEA, De Pinho SZ (2003) Germination of *Senna occidentalis* link: seed at different osmotic potential levels. Braz Arch Technol 46:163–166

Delfine S, Loreto F, Alvino A (2001) Drought-stress effects on physiology, growth and biomass production of rainfed and irrigated Bell Pepper plants in the Mediterranean region. J Am Soc Hortic Sci 126:297–304

Doussan C, Pagès L, Vercambre G (1998) Modelling of the hydraulic architecture of root systems: an integrated approach to water absorption—model description. Ann Bot 81:213–223

DTMA (2012) Drought tolerant maize for Africa: summary reports 2012. CIMMYT, Nairobi

Du Plessis J (2003) Maize production. Directorate agricultural information services, department of agriculture in cooperation with ARC-Grain Crops Institute

Dubey L, Prasanna BM, Ramesh B (2009) Analysis of drought tolerant and susceptible maize genotypes using SSR markers tagging candidate genes and consensus QTLs for drought tolerance. Indian J Genet Plant Breed 69:344–351

Duvick DN, Cassman KG (1999) Post–green revolution trends in yield potential of temperate maize in the North-Central United States. Crop Sci 39(6):1622–1630

Duvick DN, Smith JCS, Cooper M (2004) Long-term selection in a commercial hybrid maize breeding program. Plant Breed Rev 24:109–151

Eathington SR, Crosbie TM, Edwards MD, Reiter RS, Bull JK (2007) Molecular markers in a commercial breeding program. Crop Sci 47:S154–S163

Edgerton MD (2009) Increasing crop productivity to meet global needs for feed, food and fuel. Plant Physiol 149:7–13

Edmeades GO, Bolaños J, Elings A, Ribaut J-M, Bänziger M, Westgate ME (2000) The role and regulation of the anthesis-silking interval in maize. In: Westgate ME, Boote KJ (eds) Physiology and modeling kernel set in maize. CSSA Special Publication No. 29. CSSA, Madison, pp 43–73

Edmeades GO, Cooper M, Lafitte R, Zinselmeier C, Ribaut JM, Habben JE, Löffler C, Bänziger M (2001) Abiotic stresses and staple crops. In: Proceedings of the third international crop science congress, Hamburg, Germany, CABI

Edmeades GO, Bänziger M, Campos H, Schussler J (2006) Improving tolerance to abiotic stresses in staple crops: a random or planned process? In: Lamkey KR, Lee M (eds) Plant breeding: the Arnel R. Hallauer international symposium. Blackwell Publishing, Ames, pp 293–309

Edmeades GO (2008) Drought tolerance in maize: an emerging reality. In: James C (ed) ISAAA Brief 39. Global status of commercialized biotech/GM crops, ISAAA, pp 197–217

Edmeades GO (2013) Progress in achieving and delivering drought tolerance in maize—an update. Ithaca, ISAAA

Ekanayake IJ, Otoole JC, Garrity DP, Masajo TM (1985) Inheritance of root characters and their relations to drought resistance in rice. Crop Sci 25:927–933

Emam Y, Seghatoeslami MJ (2005) Crop yield, physiology and processes. Shiraz University Press, Shiraz

Falconer DS (1989) Introduction to quantitative genetics, 3rd edn. Longman, London

FAOSTAT (2013) Food and Agriculture Organization of the United Nations (FAO) Statistical Databases. http://www.fao.org/site/567/

Farooq M, Wahid A, Kobayashi N, Fujita D, Basra SMA (2009) Plant drought stress: effects, mechanisms and management. Agron Sustain Dev 29(1):185–212

Fazeli F, Ghorbanli M, Niknam V (2007) Effect of drought on biomass, protein content, lipid peroxidation and antioxidant enzymes in two sesame cultivars. Biol Plant 51:98–103

Feix G, Hochholdinger F, Wulff D (1997) Genetic analysis of root formation in maize. Developmental pathways in plants: biotechnological implications. The Hebrew University of Jerusalem, Rehovot

Feng HY, Wang ZM, Kong FN, Zhang MJ, Zhou SL (2011) Roles of carbohydrate supply and ethylene, polyamines in maize kernel set. J Integr Plant Biol 53:388–398

Fernandez GCJ (1992) Effective selection criteria for assessing plant stress tolerance. In: Proceeding of international symposium "adaptation of vegetables and other food crops in temperature and water stress". AVRDC Publisher, Tainan, pp 257–270. Accessed 13–18 Aug 1992

Finch-Savage WE (1995) Influence of seed quality on crop establishment, growth and yield. In: Basra AS (ed) Seed quality. Basic mechanisms and agricultural implications. Food products Press, New York, pp 361–384

Fischer RA, Maurer R (1978) Drought resistance in spring wheat cultivars: I. Grain yield responses. Aust J Agric Res 29:897–912

Flagella Z, Rotunno T, Tarantino E, Di Caterina R, De Caro A (2002) Changes in seed yield and oil fatty acid composition of high oleic sunflower (Helianthus annuus L.) hybrids in relation to the sowing date and the water regime. Eur J Agron 17:221–230

Flexas J, Bota J, Loreto F, Cornic G, Sharkey TD (2004) Diffusive and metabolic limitations to photosynthesis under drought and salinity in C3 plants. Plant Biol 5:1–11

Flexas J, Diaz-Espejo A, Galmés J, Kaldenhoff R, Medrano H, Ribas-Carbo M (2007) Rapid variations of mesophyll conductance in response to changes in CO_2 concentration around leaves. Plant Cell Environ 30:1284–1298

Gambín BL, Borrás L, Otegui ME (2006) Is maize kernel size limited by its capacity to expand? Mydica 52:431–441

Gavuzzi P, Rizza F, Palumbo M, Campaline RG, Ricciardi GL, Borghi B (1997) Evaluation of field and laboratory predictors of drought and heat tolerance in winter cereals. Can J Plant Sci 77:523–531

Ghannoum O (2009) C4 photosynthesis and water stress. Ann Bot 103:635–644

Gharoobi B, Ghorbani M, Nezhad MG (2012) Effects of different levels of osmotic potential on germination percentage and germination rate of barley, corn and canola. Iran J Plant Physiol 2 (2):413–417

Gilbert N (2010) Inside the hothouses of industry. Nature 466:548–551

Gill PK, Sharma AD, Singh P, Bhullar SS (2003) Changes in germination, growth and soluble sugar contents of sorghum bicolor L. Moench seeds under various abiotic stresses. Plant Growth Regul 40:157–162

Gill SS, Tuteja N (2010) Polyamines and abiotic stress tolerance in plants. Plant Signaling & Behavior 5(1):26–33

Gindaba J, Rozanov A, Negash L (2004) Response of seedlings of two Eucalyptus and three deciduous tree species from Ethiopia to severe water stress. For Ecol Manag 201:119–129

Goksoy AT, Demir AO, Turan ZM, Dagustu N (2004) Responses of sunflower to full and limited irrigation at different growth stages. Field Crops Res 87:167–178

Gomez-Roldan V, Fermas S, Brewer PB, Puech-Pages V, Dun EA, Pillot JP, Letisse F, Matusova R, Danoum S, Portais JC, Bouwmeester H, Becard G, Beveridge CA, Rameau C, Rochange SF (2008) Strigolactone inhibition of shoot branching. Nature 455:189–194

Gong H, Zhu X, Chen K, Wang S, Zhang C (2005) Silicon alleviates oxidative damage of wheat plants in pots under drought. Plant Sci 169:313–321

Granier C, Inzé D, Tardieu F (2000) Spatial distribution of cell division rate can be deduced from that of p34cdc2 kinase activity in maize leaves grown at contrasting temperatures and soil water conditions. Plant Physiol 124:1393–1402

Guei RG, Wassom CE (1993) Genetics of osmotic adjustment in breeding maize for drought tolerance. Hered 71:436–441

Guo A-Y, Chen X, Gao G, Zhang H, Zhu Q-H, Liu X-C, Zhong Y-F, Gu X, He K, Luo J (2008) PlantTFDB: a comprehensive plant transcription factor database. Nucleic Acids Res 36:D966–D969

Hadas A (2004) Seedbed preparation: the soil physical environment of germinating seeds. In: Benech-Arnold RL, Sanchez RA (eds) Handbook of seed physiology: applications to agriculture. Food Product Press, New York

Hajibabaee M, Azizi F, Zargari K (2012) Effect of drought stress on some morphological, physiological and agronomic traits in various foliage corn hybrids. Am Eurasian J Agric Environ Sci 12(7):890–896

Hallauer AR, Miranda JB (1988) Quantitative genetics in maize breeding, 2nd edn. Iowa State University Press, Ames

Hammer GL, Dong Z, McLean G, Doherty A, Messina C, Schussler J, Zinselmeier C, Paszkiewicz S, Cooper M (2009) Can change in canopy and/or root system architecture explain historical maize yield trends in the U.S. corn belt? Crop Sci 49:299–312

Hampton M, Xu WW, Kram BW, Chambers EM, Ehrnriter JS, Gralewski JH, Joyal T, Carter CJ (2010) Identification of differential gene expression in Brassica rapa nectaries through expressed sequence tag analysis. PLoS ONE 5:e8782

Harris K, Subudhi PK, Borrel A, Jordan D, Rosenow D, Nguyen H, Klein P, Klein R, Mullet J (2007) Sorghum stay-green QTL individually reduce post-flowering drought-induced leaf senescence. J Exp Bot 58:327–338

Haupt-Herting S, Fock HP (2002) Oxygen exchange in relation to carbon assimilation in water stressed leaves during photosynthesis. Ann Bot 89:851–859

Hayano-Kanashiro C, Calderón-Vázquez C, Ibarra-Laclette E, Herrera-Estrella L, Simpson J (2009) Analysis of gene expression and physiological responses in three Mexican maize landraces under drought stress and recovery irrigation. PLoS ONE 4(10):1–19

Hayat S, Ahmed A (2007) Salicylic acid: a plant hormone. Springer, The Netherland

Heinigre RW (2000) Irrigation and Drought Management. Crop Science Department. http://www.ces.ncsu.edu/plymouth/cropsci/cornguide/Chapter4.tml

Heslop-Harrison J (1979) An interpretation of the hydrodynamics of pollen. Am J Bot 66:737–743

Hospital F, Moreau L, Lacoudre F, Charcosset A, Gallais A (1997) More on the efficiency of marker-assisted selection. Theor Appl Genet 95:1181–1189

Hu H, Dai M, Yao J, Xiao B, Li X, Zhang Q, Xiong L (2006) Overexpressing a NAM, ATAF, and CUC (NAC) transcription factor enhances drought resistance and salt tolerance in rice. Proc Natl Acad Sci USA 103:12987–12992

Hu XL, Jiang MY, Zhang AY, Lin F, Tan MP (2007) Calcium/calmodulin is required for abscisic acid-induced antioxidant defense and functions both upstream and downstream of H_2O_2 production in leaves of maize plants. New Phytol 173:27–38

Hu XL, Jiang MY, Zhang JH, Tan MP, Zhang AY (2008) Cross-talk between Ca^{2+} /CaM and H_2O_2 in abscisic acid-induced antioxidant defense in leaves of maize plants exposed to water stress. Plant Grow Regul 55:183–198

Hu X, Li Y, Li C, Yang H, Wang W, Lu M (2010) Characterization of small heat shock proteins associated with maize tolerance to combined drought and heat stress. J Plant Growth Regul 29 (4):455–464

Hu X, Wu X, Li C, Lu M, Liu T, Wang Y, Wang W (2012) Abscisic acid refines the synthesis of chloroplast proteins in maize (Zea mays) in response to drought and light. PLoS ONE 7(11):1–12

Huang XY, Chao DY, Gao JP, Zhu MZ, Shi M, Lin HX (2009) A previously unknown zinc finger protein, DST, regulates drought and salt tolerance in rice via stomatal aperture control. Genes Dev 23:1805–1817

Hussain SS (2006) Molecular breeding for abiotic stress tolerance: drought perspective. Proc Pak Acad Sci 43(3):189–210

Isendahl N, Schmidt G (2006) Drought in the Mediterranean. WWF Policy Proposals, WWF Report, Madrid

Jacob T, Ritchie S, Assmann SM, Gilroy S (1999) Abscisic acid signal transduction in guard cells is mediated by phospholipase D activity. Proc Natl Acad Sci USA 96:12192–12197

Jain M, Tiwary S, Gadre R (2010) Sorbitol-induced changes in various growth and biochemical parameters in maize. Plant Soil Environ 56(6):263–267

Jaleel CA, Manivannan P, Sankar B, Kishorekumar A, Gopi R, Somasundaram R, Panneerselvam R (2007) Induction of drought stress tolerance by ketoconazole in Catharanthus roseusis mediated by enhanced antioxidant potentials and secondary metabolite accumulation. Coll Surf B: Biointerfaces 60:201–206

Janmohammadi M, Dezfuli PM, Sharifzadeh F (2008) Seed invigoration techniques to improve germination and early growth of inbred line of maize under salinity and drought stress. Gen Appl Plant Physiol 34(2–3):215–226

Jeanneau M, Gerentes D, Foueillassar X, Zivy M, Vida J, Toppan A, Perez P (2002) Improvement of drought tolerance in maize: towards the functional validation of the Zm-Asr1 gene and increase of water use efficiency by over-expressing C4–PEPC. Biochimie 84:1127–1135

Jensen ME (1973) Consumptive use of water and irrigation water requirements. American Society of Civil Engineers, New York

Jia W, Davies WJ (2007) Modification of leaf apoplastic pH in relation to stomatal sensitivity to root-sourced abscisic acid signals. Plant Physiol 143:68–77

Jones HG (1992) Plants and microclimate: a quantitative approach to environmental plant physiology, 2nd edn. Cambridge University Press, Cambridge

Jones B, Gunneras SA, Petersson SV, Tarkowski P, Graham N, May S, Dolezal K, Sandberg G, Ljung K (2010) Cytokinin regulation of auxin synthesis in Arabidopsis involves a homeostatic feedback loop regulated via auxin and cytokinin signal transduction. Plant Cell 22:2956–2969

Kakumanu A, Ambavaram MMR, Klumas C, Krishnan A, Batlang U, Myers E, Grene R, Pereira A (2012) Effects of drought on gene expression in maize reproductive and leaf meristem tissue revealed by RNA-Seq. Plant Physiol 160:846–867

Kang JY, Choi HI, Im MY, Kim SY (2002) Arabidopsis basic leucine zipper proteins that mediate stress-responsive abscisic acid signaling. Plant Cell 14:343–357

Karaba A, Dixit S, Greco R, Aharoni A, Trijatmiko KR, Marsch-Martinez N, Krishnan A, Nataraja KN, Udayakumar M, Pereira A (2007) Improvement of water use efficiency in rice by expression of HARDY, an Arabidopsis drought and salt tolerance gene. Proc Natl Acad Sci USA 104:15270–15275

Kathiresana A, Lafittea HR, Chena J, Mansuetoa L, Bruskiewicha R, Bennetta J (2006) Gene expression microarrays and their application in drought stress research. Field Crops Res 97:101–110

Kavar T, Maras M, Kidric M, Sustar-Vozlic J, Meglic V (2007) Identification of genes involved in the response of leaves of *Phaseolus vulgaris* to drought stress. Mol Breed 21:159–172

Khan MB, Hussain N, Iqbal M (2001) Effect of water stress on growth and yield components of maize variety YHS 202. J Res (Sci) 12:15–18

Khodarahmpour Z (2011) Effect of drought stress induced by polyethylene glycol (PEG) on germination indices in corn (*Zea mays* L.) hybrids. Afr J Biotechnol 10(79):18222–18227

Kiani SP, Tália P, Maury P, Grieu P, Heinz R, Perrault A, Nishinakamasu V, Hopp E, Gentzbittel L, Paniego N, Sarrafi A (2007) Genetic analysis of plant water status and osmotic adjustment in recombinant inbred lines of sunflower under two water treatments. Plant Sci 172:773–787

Kitchen NR, Sudduth KA, Drummond ST (1999) Soil electrical conductivity as a crop productivity measure for clay pan soils. J Prod Agric 12:607–617

Koornneef M, Léon-Kloosterziel KM, Schwartz SH, Zeevaart JAD (1998) The genetic and molecular dissection of abscisic acid biosynthesis and signal transduction in Arabidopsis. Plant Physiol Biochem 36:83–89

Kumar J, Abbo S (2001) Genetics of flowering time in chickpea and its bearing on productivity in the semi-arid environments. Adv Agron 72:107–138

Laporte MM, Shen B, Tarczynski MC (2002) Engineering for drought avoidance: expression of maize NADP-malic enzyme in tobacco results in altered stomatal function. J Exp Bot 53:699–705

Lauer J (2003) What happens within the corn plant when drought occurs? University of Wisconsin Extension. http://www.uwex.edu/ces/ag/issues/drought2003/corneffect.html.

Lauer J (2012) The effects of drought and poor corn pollination on corn. Field Crops 28:493–495

Lawlor DW, Cornic G (2002) Photosynthetic carbon assimilation and associated metabolism in relation to water deficits in higher plants. Plant Cell Environ 25:275–294

Leister D (2003) Chloroplast research in the genomic age. Trends Genet 19:47–56

Leitner D, Meunier F, Bodner G, Javaux M, Schnepf A (2014) Impact of contrasted maize root traits at flowering on water stress tolerance—a simulation study. Field Crops Res 15:125–137

Li L, Van Staden J, Jäger AK (1998) Effects of plant growth regulators on the antioxidant system in seedlings of two maize cultivars subjected to water stress. Plant Growth Regul 25:81–87

Li WX, Oono Y, Zhu J, He XJ, Wu JM, Iida K, Lu XY, Cui X, Jin H, Zhu JK (2008) The Arabidopsis NFYA5 transcription factor is regulated transcriptionally and post transcriptionally to promote drought resistance. Plant Cell 20:2238–2251

Liu J, Jiang MY, Zhou YF, Liu YL (2005) Production of polyamines is enhanced by endogenous abscisic acid in maize seedlings subjected to salt stress. J Integr Plant Biol 47:1326–1334

Liu XA, Bush DR (2006) Expression and transcriptional regulation of amino acid transporters in plants. Amino Acids 30(2):113–120

Lorenz AJ, Chao S, Asoro FS, Heffner EL, Hayashi T, Iwata H, Smith KP, Sorrells ME, Jannink J-L (2011) Genomic selection in plant breeding: knowledge and prospects. Adv Agron 110:77–123

Lu GH, Tang JH, Yan JB, Ma XQ, Li JS, Chen SJ, Ma JC, Liu ZX, LZ E, Zhang YR, Dai JR (2006) Quantitative trait loci mapping of maize yield and its components under different water treatments at flowering time. J Integr Plant Biol 48(10):1233–1243

Lü B, Gong Z, Wang J, Zhang J, Liang J (2007) Microtubule dynamics in relation to osmotic stress-induced ABA accumulation in Zea mays roots. J Exp Bot 58(10):2565–2572

Ludlow MM, Muchow RC (1990) A critical evaluation of traits for improving crop yields in water-limited environments. Adv Agron 43:107–153

Mabhaudhi T (2009) Response of maize (Zea mays L.) landraces to water stress compared with commercial hybrids. Crop Science, School of Agricultural Sciences and Agribusiness, Faculty of Science and Agriculture, University of KwaZulu-Natal, P/Bag X01, Scottsville, 3209, Pietermaritzburg, South Africa

Makela P, Karkkainen J, Somersalo S (2000) Effect of glycine betaine on chloroplast ultrastructure, chlorophyll and protein content, and RuBPCO activity in tomato grown under drought or salinity. Biol Plant 43:471–475

Mambelli S, Setter TL (1998) Inhibition of maize endosperm cell division and endoreduplication by exogenous applied abscisic acid. Physiol Plant 104:266–277

Marcus AI, Moore RC, Cyr RJ (2001) The role of microtubules in guard cell function. Plant Physiol 125:387–395

Marivet J, Frendo P, Burkard G (1992) Effects of abiotic stresses on cyclophilin gene expression in maize and bean and sequence analysis of bean cyclophilin cDNA. Plant Sci 84(2):171–178

McCue KF, Hanson AD (1990) Drought and salt tolerance: towards understanding and application. Trends Biotech 8:358–362

Medrano H, Escalona JM, Bota J, Gulias J, Flexas J (2002) Regulation of photosynthesis of C3 plants in response to progressive drought. Stomatal conductance as a reference parameter. Ann Bot 89:895–905

Meuwissen T, Hayes BJ, Goddard ME (2001) Prediction of total genetic value using genome wide dense marker maps. Genet 157:1819–1829

Mir RR, Zaman-Allah M, Sreenivasulu N, Trethowan R, Varshney RK (2012) Integrated genomics, physiology and breeding approaches for improving drought tolerance in crops. Theor Appl Genet 125(4):625–645

Mizoguchi T, Ichimura K, Yoshida R, Shinozaki K (2000) MAP kinase cascades in Arabidopsis: their roles in stress and hormone responses. Results Probl Cell Differ 27:29–38

Mohammadkhani N, Heidari R (2008) Drought-induced accumulation of soluble sugars and proline in two maize varieties. World Appl Sci J 3(3):448–453

Moose SP, Lauter N, Carlson SR (2004) The maize macrohairless1 locus specifically promotes leaf blade macrohair initiation and responds to factors regulating leaf identity. Genetics 166:1451–1461

Morgan J, Hare R, Fletcher R (1986) Genetic variation in osmoregulation in bread and durum wheats and its relationship to grain yield in a range of field environments. Crop Pasture Sci 37:449–457

Moss GI, Downey LA (1971) Influence of drought stress on female gametophyte development in corn (Zea mays L.) and subsequent grain yield. Crop Sci 11:368–372

Mundy J, Yamaguchi-Shinozaki K, Chua N-H (1990) Nuclear proteins bind conserved elements in the abscisic acid-responsive promoter of a rice Rab gene. Proc Natl Acad Sci USA 87:1406–1410

Nakashima K, Tran LS, Van Nguyen D, Fujita M, Maruyama K, Todaka D, Ito Y, Hayashi N, Shinozaki K, Yamaguchi-Shinozaki K (2007) Functional analysis of a NAC-type transcription factor OsNAC6 involved in abiotic and biotic stress-responsive gene expression in rice. Plant J 51:617–630

Navara L, Jeˇsko T, Duchoslav S (1994) Participation of seminal roots in water uptake by maize root system. Biologia (Bratislava) 49:91–95

Naveed S, Aslam M, Maqbool MA, Bano S, Zaman QU, Ahmad RM (2014) Physiology of hightemperature stress tolerance at reproductive stages in maize. J Anim Plant Sci 24(4):1141–1145

Nejad SK, Bakhshande A, Nasab SB, Payande K (2010) Effect of drought stress on corn root growth. Rep Opin 2(2):1–7

Nelson DE, Repetti PP, Adams TR, Creelman RA, Wu J, Warner DC, Anstrom DC, Bensen RJ, Castiglioni PP, Donnarummo MG, Hinchey BS, Kumimoto RW, Maszle DR, Canales RD, Krolikowski KA, Dotson SB, Gutterson N, Ratcliffe OJ, Heard JE (2007) Plant nuclear factor Y (NF-Y) B subunits confer drought tolerance and lead to improved corn yields on water-limited acres. Proc Natl Acad Sci USA 104:16450–16455

Neuffer M, Coe E, Wessler SR (1997) Mutants in maize. Cold Spring Harbor Laboratory Press, Cold Spring Harbor

Nguyen TX, Nguyen T, Alameldin H, Goheen B, Loescher W, Sticklen M (2013) Transgene pyramiding of the *HVA1* and *mtlD* in T3 maize (*Zea mays* L.) plants confers drought and salt tolerance, along with an increase in crop biomass. Int J Agron 2013:1–10

Nielsen RL (2002) A fast and accurate pregnancy test for corn. Chat'n chew cafe. http://www.kingcorn.org/news/articles.02/Pregnancy_Test-0717.html.

Nielsen RL (Bob) (2005a) Silk emergence. Corny News Network, Purdue University. http://www.agry.purdue.edu.

Nielsen RL (Bob) (2005b) Tassel emergence and pollen shed. Corny News Network, Purdue University. http://www.agry.purdue.edu.

Nonami H (1998) Plant water relations and control of cell elongation at low water potentials. J Plant Res 111:373–382

Norastehnia A, Sajedi RH, Nojavan-Asghari M (2007) Inhibitory effects of methyl jasmonate on seed germination in maize (*Zea Mays* L.): effect on amylase activity and ethylene production. Gen Appl Plant Physiol 33:13–23

Ober ES, Setter TL, Madison JT, Thompson JF, Shapiro PS (1991) Influence of water deficit on maize endosperm development. Enzymes activities and RNA transcripts of starch and zein synthesis, abscisic acid, and cell division. Plant Physiol 97:154–164

Oh SJ, Song SI, Kim YS, Jang HJ, Kim SY, Kim M, Kim YK, Nahm BH, Kim JK (2005) Arabidopsis CBF3/DREB1A and ABF3 in transgenic rice increased tolerance to abiotic stress without stunting growth. Plant Physiol 138:341–351

Olsen O-A, Linnestad C, Nichols SE (1999) Developmental biology of the cereal endosperm. Trends Plant Sci 4:253–257

Oualset CO (1979) Breeding for drought resistance in maize. In: Proceedings of the SAFGRAD/International institute for tropical maize workshop. Ougadougou, Upper Volta

Oveysi M, Mirhadi MJ, Madani H, Nourmohammadi G, Zarghami R, Madani A (2010) The impact of source restriction on yield formation of corn (*Zea mays* L.) due to water deficiency. Plant Soil Environ 56(10):476–481

Pannar (2012). Quality seed, know the maize plant. Pannar Seeds, Private Limited

Parry MAJ, Flexas J, Medrano H (2005) Prospects for crop production under drought: research priorities and future directions. Ann Appl Biol 147:211–226

Passioura JB (1996) Drought and drought tolerance. Plant Growth Reg 20(2):79–83

Passioura JB (2007) The drought environment: physical, biological and agricultural perspectives. J Exp Bot 58(2):113–117

Phelps TL, Hall AE, Buckner B (1996) Microsatellites repeat variation within the y1 gene of maize and teosinte. J Hered 87:396–399

Pirasteh-Anosheh H, Emam Y, Pessarakli M (2013) Changes in endogenous hormonal status in corn (*Zea Mays* L.) Hybrids under drought stress. J Plant Nutr 36(11):1695–1707

Podlich DW, Winkler CR, Cooper M (2004) Mapping as you go: an effective approach for marker-assisted selection of complex traits. Crop Sci 44:1560–1571

Pollock FM, Pickett-Heaps JD (2005) Spatial determinants in morphogenesis: recovery from plasmolysis in the diatom Ditylum. Cell Motil Cytoskelet 60:71–82

Qin F, Sakuma Y, Li J, Liu Q, Li Y, Shinozaki K, Yamaguchi-Shinozaki K (2004) Cloning and functional analysis of a novel DREB1/CBF transcription factor involved in cold responsive gene expression in *Zea mays* L. Plant and Cell Physiol 45:1042–1052

Qin L, Trouverie J, Chateau-Joubert S, Simond-Cote E, Thevenot C, Prioul J-L (2004) Involvement of the Ivr2-invertase in the perianth during maize kernel development under water stress. Plant Sci 166:371–379

Qin F, Kakimoto M, Sakuma Y, Maruyama K, Osakabe Y, Phan Tran L-S, Shinozaki K, Yamaguchi-Shinozak K (2007) Regulation and functional analysis of ZmDREB2A in response to drought and heat stresses in Zea mays L. Plant J 50:54–69

Qualset CO (1979) Breeding for drought resistance in maize. In: Proceedings of the SAFGRAD/Internatiorial institute for tropical maize workshop. Ougadougou, Upper Volta

Quan R, Shang M, Zhang H, Zhao Y, Zhang J (2004) Engineering of enhanced glycine betaine synthesis improves drought tolerance in maize. Plant Biotechnol J 2(6):477–486

Radić V, Vujaković M, Jeromela AM (2007) Influence of drought on seedling development in different corn genotypes (Zea mays L.). J Agric Sci 52(2):131–136

Ragot M, Biasiolli M, Delbut MF, Dell'Orco A, Malgarini L, Thevenin P, Vernoy J, Vivant J, Zimmermann R, Gay G (1995) Marker-assisted backcrossing: a practical example. In: Berville A, Tersac M (eds) Les Colloques, No 72, Techniques reutilizations des marqueurs moléculaires, Paris, INRA, pp 45–56

Rana BS, Rao MH (2000) Technology for increasing sorghum production and value addition. National Research Center for Sorghum, Indian Council of Agricultural Research, Hyderabad, India

Rao SR, Qayyum A, Razzaq A, Ahmad M, Mahmood I, Sher A (2012) Role of foliar application of salicylic acid and l-tryptophan in drought tolerance of maize. J Anim Plant Sci 22:768–772

Rauf M, Munir M, Hassan M, Ahmad M, Afzal M (2007) Performance of wheat genotypes under osmotic stress at germination and early seedling growth stage. Afr J Biotech 6:971–975

Rea G, de Pinto MC, Tavazza R, Biondi S, Gobbi V, Ferrante P, Gara LD, Federico R, Angelini R, Tavladoraki P (2004) Ectopic expression of maize polyamine oxidase and pea copper amine oxidase in the cell wall of tobacco plants. Plant Physiol 134(4):1414–1426

Ren C, Han C, Peng W, Huang Y, Peng Z, Xiong X, Zhu Q, Gao B, Xie D (2009) A leaky mutation in DWARF4 reveals an antagonistic role of brassinosteroid in the inhibition of root growth by jasmonate in Arabidopsis. Plant Physiol 151:1412–1420

Reynolds M, Tuberosa R (2008) Translational research impacting on crop productivity in drought-prone environments. Curr Opin Plant Biol 11:171–179

Rhoads FM, Bennett JM (1990) Corn. In: Stewart BA, Nielsen DR (eds) Irrigation of agricultural crops. ASA-CSSA-SSSA, Madison

Ribaut J-M, Ragot M (2007) Marker-assisted selection to improve drought adaptation in maize: the backcross approach, perspectives, limitations, and alternatives. J Exp Bot 58(2):351–360

Richards RA, Rawson HM, Johnson DA (1986) Glaucousness in wheat: its development, and effect on water-use efficiency, gas exchange and photosynthetic tissue temperatures. Aust J Plant Physiol 13:465–473

Ristic Z, Gifford DJ, Cass DD (1991) Heat shock proteins in two lines of Zea mays L. That differ in drought and heat resistance. Plant Physiol 97:1430–1434

Rosielle AA, Hambling J (1981) Theoretical aspects of selection for yield in stress and non-stress environments. Crop Sci 21:943–946

Roychowdhury R, Tah J (2013) Mutagenesis-a potential approach for crop improvement. In: Hakeem KR, Ahmad P, Ozturk M (eds) Crop improvement. Springer, Berlin

Rucker KS, Kvien CK, Holbrook CC, Hook JE (1995) Identification of peanut genotypes with improved drought avoidance traits. Peanut Sci 24:14–18

Rutkoski JE, Heffner EL, Sorrells ME (2010) Genomic selection for durable stem rust resistance in wheat. Euphytica 179:161–173

Sage RF, Zhu XG (2011) Exploiting the engine of C4 photosynthesis. Exp Bot 62(9):2989–3000

Saini HS, Westgate ME (2000) Reproductive development in grain crops during drought. Adv Agron 68:59–96

Sakamoto H, Araki T, Meshi T, Iwabuchi M (2000) Expression of a subset of the Arabidopsis Cys (2)/His(2)-type zinc-finger protein gene family under water stress. Gene 248:23–32

Sakamoto H, Maruyama K, Sakuma Y, Meshi T, Iwabuchi M, Shinozaki K, Yamaguchi Shinozaki K (2004) Arabidopsis Cys2/His2-type zinc-finger proteins function as transcription repressors under drought, cold, and high-salinity stress conditions. Plant Physiol 136:2734–2746

Sanchez JP, Chua NH (2001) Arabidopsis PLC1 is required for secondary responses to abscisic acid signals. Plant Cell 13:1143–1154

Sandquist DR, Ehleringer JR (2003) Population- and family-level variation of brittlebush (*Encelia farinosa*, Asteraceae) pubescence: its relation to drought and implications for selection in variable environments. Am J Bot 90:1481–1486

Sang Y, Cui D, Wang X (2001) Phospholipase D and phosphatidic acid—mediated generation of superoxide in Arabidopsis. Plant Physiol 126:1449–1458

Sawkins M, Meyer J, Ribaut JM (2006) Drought adaptation in cereal crops drought adaptation in maize. In: Ribaut JM (ed) Drought tolerance in cereals. The Haworth Press Inc, Binghamtown, pp 356–387

Schoper JB, Lambert RJ, Vasilas BL, Westgate ME (1987) Plant factors controlling seed set in maize: the influence of silk, pollen, and ear-leaf water status and tassel heat treatment at pollination. Plant Physiol 83:121–125

Schussler J, Barker T, Lee T, Loffler C, Hausmann N, Keaschall J, Paszkiewicz S, Leafgren R, Gho C, Cooper M (2011) Development of commercial drought tolerant maize hybrids. Poster presented at CSSA annual meetings, October, 2011

Sebastian S (2009) Accelerated yield technology: context-specific MAS for grain yield. http://www.plantsciences.ucdavis.edu. Accessed 19 Nov 2012

Semagn K, Bjørnstad Å, Ndjiondjop MN (2006) An overview of molecular marker methods for plants. Afr J Biotechnol 5:2540–2568

Serraj R, Hash TC, Buhariwalla HK, Bidinger FR, Folkertsma RT, Chandra S, Gaur PM, Kashiwagi J, Nigam SN, Rupakula A, Crouch JH (2005a) Marker-assisted breeding for crop drought tolerance at ICRISAT: achievements and prospects. In: Tuberosa R, Phillips RL, Gale M (eds) Proceedings of the international congress "In the wake of the double helix: from the green revolution to the gene revolution". Avenue Media, Bologna, pp 217–238

Serraj R, Hash CT, Rizvi SM, Sharma A, Yadav RS, Bidinger FR (2005b) Recent advances in marker assisted selection for drought tolerance in pearl millet. Plant Prod Sci 8:334–337

Setter TL, Flannigan BA (2001) Water deficit inhibits cell division and expression of transcripts involved in cell proliferation and endoreduplication in maize endosperm. J Exp Bot 52 (360):1401–1408

Setter TL, Flannigan BA, Melkonian J (2001) Loss of kernel set due to water deficit and shade in maize: carbohydrate supplies, abscisic acid, and cytokinins. Crop Sci 41:1530–1540

Setter TL, Yan J, Warburton M, Ribaut JM; Xu Y, Sawkins M, Buckler ES, Zhang Z, Gore MA (2011) Genetic association mapping identifies single nucleotide polymorphisms in genes that affect abscisic acid levels in maize floral tissues during drought. J Exp Bot 62:701–716

Shaddad MAK, El-Samad MHA, Mohammed HT (2011) Interactive effects of drought stress and phytohormones or polyamines on growth and yield of two M (*Zea maize* L.) genotypes. Am J Plant Sci 2:790–807

Shao HB, Shao MA, Liang ZS (2006) Osmotic adjustment comparison of 10 wheat (Triticum aestivum L.) genotypes at soil water deficits. Coll Surf B: Biointerfaces 47(2):132–139

Shaw RH (1976) Secondary wind maxima inside plant canopies. J Appl Meteorol 16:514–521

Sheen J (1996) Ca^{2+} dependent protein kinases and stress signal transduction in plants. Science 274:1900–1902

Shen YY, Wang XF, Wu FQ, Du SY, Cao Z, Shang Y, Wang XL, Peng CC, Yu XC, Zhu SY, Fan RC, Xu YH, Zhang DP (2006) The Mg-chelatase H subunit is an abscisic acid receptor. Nature 443:823–826

Sheoran IS, Saini HS (1996) Drought-induced sterility in rice: changes in carbohydrate levels and enzyme activities associated with the inhibition of starch accumulation in pollen. Sex Plant Reprod 9:1661–1669

Shinozaki K, Yamaguchi-Shinozaki K (1997) Gene expression and signal transduction in water stress response. Plant Physiol 115:327–334

Shinozaki K, Yamaguchi-Shinozaki K, Seki M (2003) Regulatory network of gene expression in the drought and cold stress responses. Curr Opin Plant Biol 6:410–417

Shou H, Bordallo P, Wang K (2004) Expression of the Nicotiana protein kinase (NPK1) enhanced drought tolerance in transgenic maize. J Exp Bot 55(399):1013–1019

Siddique MRB, Hamid A, Islam MS (2001) Drought stress effects on water relations of wheat. Bot Bull Acad Sin 41:35–39

Singh RK, Chaudhary BD (1985) Biometrical methods in quantitative genetics analysis. Kalyani Publishers, New Delhi

Singh BD (2010) Plant breeding: principles and methods, 8th edn. Kalyani Publishers, New Dehli

Sreenivasulu N, Sopory SK, Kishor PBK (2007) Deciphering the regulatory mechanisms of abiotic stress tolerance in plants by genomic approaches. Gene 388:1–13

Steele KA, Virk DS, Kumar R, Prasad SC, Witcombe JR (2007) Field evaluation of upland rice lines selected for QTLs controlling root traits. Field Crops Res 101:180–186

Steinborn K, Maulbetsch C, Priester B, Trautmann S, Pacher T, Geiges B, Küttner F, Lepiniec L, Stierhof YD, Schwarz H, Jürgens G, Mayer U (2002) The Arabidopsis PILZ group genes encode tubulin-folding cofactor orthologs required for cell division but not cell growth. Genes Dev 16:959–971

Stoskopf CN (1981) Understanding crop production. Reston Publishing Company Inc, Reston, pp 91–99

Subbarao GV, Nam NH, Chauhan YS, Johansen C (2000) Osmotic adjustment, water relations and carbohydrate remobilization in pigeonpea under water deficits. J Plant Physiol 157:651–659

Subbarao GV, Ito O, Serraj R, Crouch JJ, Tobita S, Okada K, Hash CT, Ortiz R, Berry WL (2005) Physiological perspectives on improving crop adaptation to drought-justification for a systematic component-based approach. In: Pessarakli M (ed) Handbook of photosynthesis, 2nd edn. Marcel and Dekker, New York, pp 577–594

Syvänen AC (2005) Toward genome-wide SNP genotyping. Nat Genet 37:S5–S10

Tai FJ, Yuan ZL, Wu XL, Zhao PF, Hu XL, Wang W (2011) Identification of membrane proteins in maize leaves, altered in expression under drought stress through polyethylene glycol treatment. Plant Omics J 4:250–256

Taiz L, Zeiger E (2006) Stress physiology. In: Taiz L, Zeiger E (eds) Plant physiology. Sinauer Associates Inc, Sunderland

Taji T, Ohsumi C, Iuchi S, Seki M, Kasuga M, Kobayashi M, Yamaguchi Shinozaki K, Shinozaki K (2002) Important role of drought and cold inducible genes for galactinol synthase in stress tolerance in Arabidopsis thaliana. Plant J 29:417–426

Tanaka A, Nakagawa H, Tomita C, Shimatani Z, Ohtake M, Nomura T, Jiang C-J, Dubouzet JG, Kikuchi S, Sekimoto H, Yokota T, Asami T, Kamakura T, Mori M (2009) Brassinosteroid Upregulated1, encoding a helix–loop–helix protein, is a novel gene involved in brassinosteroid signaling and controls bending of the lamina joint in rice. Plant Physiol 151:669–680

Tanguilig VC, Yambao EB, Toole JC, Datta SKD (1987) Water stress effects on leaf elongation, leaf water potential, transpiration, and nutrient uptake of rice, maize, and soybean. Plant Soil 103:155–168

Tardieu F, Reymond M, Hamard P, Granier C, Muller B (2000) Spatial distribution of expansion rate, cell division rate and cell size in maize leaves: a synthesis of the effects of soil water status, evaporation demand and temperature. J Exp Bot 51(350):1505–1514

Tezara W, Mitchell VJ, Driscoll SD, Lawlor DW (1999) Water stress inhibits plant photosynthesis by decreasing coupling factor and ATP. Nature 401:914–917

Thitamadee S, Tuchihara K, Hashimoto T (2002) Microtubule basis for left-handed helical growth in Arabidopsis. Nature 417:193–196

Throneberry GO, Smith FG (1955) Relation of respiratory and enzymatic activity to corn seed viability. Plant Physiol 30:337–343

Timson J (1965) New method of recording germination data. Nature 207:216–217

Tollefson J (2011) Drought-tolerant maize gets US debut. Nature 469:144

Tollenaar M, Lee EA (2011) Strategies for enhancing grain yield in maize. Plant Breed Rev 34:37–82

Tran LS, Nakashima K, Sakuma Y, Simpson SD, Fujita Y, Maruyama K, Fujita M, Seki M, Shinozaki K, Yamaguchi-Shinozaki K (2004) Isolation and functional analysis of Arabidopsis stress-inducible NAC transcription factors that bind to a drought responsive cis-element in the early responsive to dehydration stress1 promoter. Plant Cell 16:2481–2498

Trivedi DK, Ansari MW, Tuteja N (2013) Multiple abiotic stress responsive rice cyclophilin: (OsCYP-25) mediates a wide range of cellular responses. Commun Integr Biol 6(5):e25260–e25268

Trivedi DK, Yadav S, Vaid N, Tuteja N (2012) Genome wide analysis of Cyclophilin gene family from rice and Arabidopsis and its comparison with yeast. Plant Signal Behav 7:1653–1666

Tsuchisaka A, Theologis A (2004) Unique and overlapping expression patterns among the Arabidopsis 1-amino-cyclopropane-1-carboxylate synthase gene family members. Plant Physiol 136:2982–3000

Tuberosa R, Salvi S, Sanguineti MC, Landi P, Maccaferri M, Conti S (2002a) Mapping QTLs regulating morpho-physiological traits and yield: case studies, shortcomings and perspectives in drought-stressed maize. Ann Bot 89:941–963

Tuberosa R, Sanguineti MC, Landi P, Giuliani MM, Salvi S, Conti S (2002b) Identification of QTLs for root characteristics in maize grown in hydroponics and analysis of their overlap with QTLs for grain yield in the field at two water regimes. Plant Mol Biol 48:697–712

Turner NC, Wright GC, Siddique KHM (2001) Adaptation of grain legumes (pulses) to water limited environments. Adv Agron 71:123–231

Ueda J, Saniewski J (2006) Methyl jasmonate-induced stimulation of chlorophyll formation in the basal part of tulip bulbs kept under natural light conditions. J Fruit Ornam Plant Res 14:199–210

Varshney RK (2010) Gene-based marker systems in plants: high throughput approaches for discovery and genotyping. In: Jain SM, Brar DS (eds) Molecular techniques in crop improvement. Springer, The Netherlands, pp 119–142

Vinocur B, Altman A (2005) Recent advances in engineering plant tolerance to abiotic stress: achievements and limitations. Curr Opin Biotechnol 16:123–132

Vivek BS (2013) Principal investigator, SP3 PROJECT G4008.56, Asian Maize Drought Tolerance (AMDROUT) Project

Von Caemmerer S (2000) Biochemical models of leaf photosynthesis. CSIRO Publishing, Collingwood

Wahid A, Gelani S, Ashraf M, Foolad MR (2007) Heat tolerance in plants: an overview. Environ Exp Bot 61:199–223

Wang W, Vinocur B, Altman A (2003) Plant responses to drought, salinity and extreme temperatures: towards genetic engineering for stress tolerance. Planta 218:1–14

Wang C, Yang A, Yin H, Zhang J (2008) Influence of water stress on endogenous hormone concentrations and cell damage of maize seedlings. Integr Plant Bio 50:427–434

Waugh R, Leader DJ, McCallum N, Caldwell D (2006) Harvesting the potential of induced biological diversity. Trends Plant Sci 11:71–79

Weiss D, Ori N (2007) Mechanisms of cross talk between gibberellin and other hormones. Plant Physiol 144:1240–1246

Wesley B, Bruce Edmeades GO, Barker TC (2002) Molecular and physiological approaches to maize improvement for drought tolerance. J Exp Bot 53(366):13–25

Westgate ME, Otegui ME, Andrade FH (2004) Physiology of the corn plant. In: Smith WC, Betrán J, Runge ECA (ed) Corn: origin, history, technology and production. Wiley, Hoboken, pp 235–271

Whittington AT, Vugrek O, Wei KJ, Hasenbein NG, Sugimoto K, Rashbrooke MC, Wasteneys GO (2001) MOR1 is essential for organizing cortical microtubules in plants. Nature 411:610–613

Wilde HD, Chen Y, Jiang P, Bhattacharya A (2012) Targeted mutation breeding of horticultural plants. Emir J Food Agric 24(1):31–41

Wilkinson S, Davies WJ (2010) Drought, ozone, ABA and ethylene: new insights from cell to plant to community. Plant Cell Environ 33:510–525

Wilkinson S, Kudoyarova GR, Veselov DS, Arkhipova TN, Davies WJ (2012) Plant hormone interactions: innovative targets for crop breeding and management. J Exp Bot 63(9):3499–3509

Williamson JD, Stoop JM, Massel MO, Conkling MA, Pharr DM (1995) Sequence analysis of a mannitol dehydrogenase cDNA from plants reveals a function for the pathogenesis related protein ELI3. Proc Natl Acad Sci USA 92:7148–7152

Wu Y, Kuzma J, Marechal E, Graeff R, Lee HC, Foster R, Chua NH (1997) Abscisic acid signaling through cyclic ADP-ribose in plants. Sci 278:2126–2130

Xia XJ, Wang YJ, Zhou YH, Tao Y, Mao WH, Shi K, Asami T, Chen Z, Yu J-Q (2009) Reactive oxygen species are involved in brassinosteroid-induced stress tolerance in cucumber. Plant Physiol 150:801–814

Xiao BZ, Chen X, Xiang CB, Tang N, Zhang QF, Xiong LZ (2009) Evaluation of seven function-known candidate genes for their effects on improving drought resistance of transgenic rice under field conditions. Mol Plant 2:73–83

Xiong L, Ishitani M, Lee H, Zhu JK (2001) The Arabidopsis *LOS5/ABA3* locus encodes a molybdenum cofactor sulfurase and modulates cold and osmotic stress responsive gene expression. Plant Cell 13:2063–2083

Xu DQ, Huang J, Guo SQ, Yang X, Bao Y-M, Tang H-J, Zhang H-S (2008) Overexpression of a TFIIIA-type zinc finger protein gene ZFP252 enhances drought and salt tolerance in rice (*Oryza sativa* L.). FEBS Lett 582:1037–1043

Xu Y, Skinner DJ, Wu H, Palacios-Rojas N, Araus JL, Yan J, Gao S, Warburton ML, Crouch JH (2009) Advances in maize genomics and their value for enhancing genetic gains from breeding. Int J Plant Genomics 2009:1–30

Yadav RS, Hash CT, Bidinger FR, Devos KM, Howarth CJ (2004) Genomic regions associated with grain yield and aspects of post flowering drought tolerance in pearl millet across environments and tester background. Euphytica 136:265–277

Yamaguchi-Shinozaki K, Shinozaki K (2006) Transcriptional regulatory networks in cellular responses and tolerance to dehydration and cold stresses. Annu Rev Plant Biol 57:781–803

Yan J, Warburton M, Crouch J (2011) Association mapping for enhancing maize (*Zea mays* L.) genetic improvement. Crop Sci 51:433–449

Yancey PH (1994) Compatible and counteracting solutes. In: Strange K (ed) Cellular and molecular physiology of cell volume regulation. CRC Press, Boca Raton, pp 81–109

Yang J, Zhang J, Wang Z, Xu G, Zhu Q (2004) Activities of key enzymes in sucrose-to starch conversion in wheat grains subjected to water deficit during grain filling. Plant Physiol 135:1621–1629

Yang S, Vanderbeld B, Wan J, Huang Y (2010) Narrowing down the targets: towards successful genetic engineering of drought-tolerant crops. Mol Plant 3(3):469–490

Zeid IM, Nermin AE (2001) Responses of drought tolerant varieties of maize to drought stress. Pak J Biol Sci 4:779–784

Zhang A, Zhang J, Zhang J, Ye N, Zhang H, Tan M, Jiang M (2011) Nitric oxide mediates brassinosteroid-induced ABA biosynthesis involved in oxidative stress tolerance in maize leaves. Plant Cell Physiol 52(1):181–192

Zhang J, Jia W, Yang J, Ismail AM (2006) Role of ABA in integrating plant responses to drought and salt stresses. Field Crops Res 97:111–119

Zhao TJ, Sun S, Liu Y, Liu JM, Liu Q, Yan YB, Zhou HM (2006) Regulating the drought responsive element (DRE)-mediated signaling pathway by synergic functions of trans active and transinactive DRE binding factors in *Brassica napus*. J Biol Chem 281:10752–10759

Zhou L, Vandersteen J, Wang L, Fuller T, Taylor M, Palais B, Wittwer CT (2004) High resolution DNA melting curve analysis to establish HLA genotypic identity. Tissue Antigens 64:156–164

Zhou LM, Wang L, Palais R, Pryor R, Wittwer CT (2005) High resolution DNA melting analysis for simultaneous mutation scanning and genotyping in solution. Clin Chem 51:1770–1777

Zhu JK (2002) Plant salt tolerance. Trends Plant Sci 6:66–71

Zhu JK, Hasegawa PM, Bressan RA (1996) Molecular aspects of osmotic stress in plants. Crit Rev
   Plant Sci 16:253–277
Zhuang Y, Ren G, Yue G, Li Z, Qu X, Hou G, Zhu Y, Zhang J (2007) Effects of water deficit
   stress on the transcriptomes of developing immature ear and tassel in maize. Plant Cell Rep
   26:2137–2147
Zinselmeier C, Jeong BR, Boyer JS (1999) Starch and the control of kernel number in maize at low
   water potentials. Plant Physiol 121:25–36
Zinselmeier C, Sun Y, Helentjaris T, Beatty M, Yang S, Smith H, Habben J (2002) The use of
   gene expression profiling to dissect the stress sensitivity of reproductive development in maize.
   Field Crops Res 75:111–121

Printed in the United States
By Bookmasters